rain & shine

An essential primer on how crop plants grow to feed the world

DAVID F SMITH

Revised edition

DUNCAN A ROUCH

Published in 2019 by Connor Court Publishing Pty Ltd

Connor Court Publishing Pty Ltd.

PO Box 7257

Redlands Bay Qld 4165

www.connorcourtpublishing.com.au

sales@connorcourt.com

ISBN: 9781925826388

Front cover design: Maria Giordano

Front cover picture: Pic by Neil Palmer (CIAT). A farmer at work in Kenya's Mount Kenya region, Wikipedia Commons.

Printed and bound in Australia

"This book is dedicated to the memory of the late Dr David Smith, who passionately devoted his professional life to supporting sustainable agriculture. In part, he loved to teach students about fundamental science and its application to agriculture, along with the related global challenges. This book is based on his deep expert knowledge and experience, developed over his life time."

Contents

List of topics

8 From gardening to farming

List of images

Glossary of key terms

Bare fallow land: land kept nearly free of any growth.

Break in the season: when soil moisture becomes available, at the start of the growing season.

Conservation farming: operations that combine leaving all plant residues on the surface of the soil, and in many cases minimum tillage (min-till).

Cultivar: an agricultural or horticultural variety or strain of a plant that originated, and is persistent, under cultivation.

Drought: a comparative rather than an absolute term. For agriculture it is a longer-than-usual period of lower-than-usual rain at that place, so that growth of plants is reduced or their survival threatened.

Effective rainfall: the part of rainfall used by plants.

Evaporation: the transfer of water from soil to the atmosphere.

Evapotranspiration (ET): = transpiration + evaporation.

Fallow land: land that was out of crop sequence, resting, and used for animal grazing.

Field Capacity: The moisture content of soil in the field as measured two or three days after a thorough wetting of a well-drained soil by rain or irrigation water.

Fertilizers: Any organic or inorganic material of natural or synthetic origin which is added to soil to provide nutrients, including nitrogen, phosphorus, and potassium, necessary to sustain plant growth.

GPS: Geographic Positioning System.

Growing season: the time period through the year when there has been moisture in the soil available to the plants, and temperature is not adverse for growth.

Peasant farming: unit areas of cultivation are small and common elements have been low level of capital investment, little mechanisation, high proportion of human physical effort, low inputs of chemicals, poor yields, and the absence of scientific support.

Permanent Wilting Point: The point at which a plant is dried so badly that even though put into a humid atmosphere and watered, it will no longer recover.

Photosynthesis: the production of a carbohydrate, glucose, in the presence of sunlight.

Phenology: the study of plants at different growth stages.

Respiration: The oxidation of carbohydrates, like glucose, which provides the release of chemical energy to maintain the metabolic processes of living things.

Senescence: Many crop plants, in the later stages, divert all products to seed, so without new leaves to maintain the increasing demand for respiration, they die.

Slash-and-burn agriculture: A variant of peasant farming, usually for corn production.

Sustainable agriculture: Agriculture to which society has committed enough resources to monitor the system, research its problems, and ensure fixing them.

Traditional agriculture: Farming based on the long-used practices, things that have been part of farming for centuries. That is, deep human involvement in processes, so less mechanisation, minimal use of chemical solutions to control pests, so, often lower yields per unit area, less use of so-called artificial fertilisers, so, at best steady yields, but often declining yields.

Transpiration: the transfer of water from plants, from pores in the leaves, to the atmosphere.

Water holding capacity: Total amount of water held in the soil per increment of depth. It is the amount of water held between field capacity and oven dry moisture level.

Wilting Point: The soil water content below which plants growing in that soil will remain wilted even when transpiration is nearly eliminated. For many agricultural soils, this represents a loss of about half the total water-holding capacity of a soil.

Dear Reader

It has never been more important than now to understand how to optimise the growth of crops plants in agriculture. As a result of continued increases in the human population on planet earth, and changes in consumer behaviour, agricultural production needs to increase by 60% over the next 40 years to meet the rising food demand. People in developing countries are gaining increased incomes, which is resulting in additional incomes being spent to purchase foods, rather than growing food at home. Total arable land is projected to increase only by less than 5% by 2050, so additional production will need to come from increased productivity.

Agriculture also, provides livelihoods for almost two-thirds of the world's extremely poor, that is, some 750 million people. This is a further substantial challenge to improving productivity while maintaining sustainability.

This book, a primer, aims to refresh your understanding of the fairly simple story of what makes plants grow, and its place in modern agriculture. We are aware that you may have been acquainted with some key parts of the story at high school, but the rest may remain a mystery! We are here to help you fill the gaps and to put the jig-saw together to gain an integrated working knowledge of what makes plants grow. We hope this book will give you unified knowledge of the great miracle of the green plant that has transformed the planet Accordingly, we have tried to explain all central concepts as simply as possible, linking these together, keeping the text short enough to be scanned at a few sittings, which you can then retain for easy reference.

After the spark of life itself, the growth of the green plant is arguably the next greatest miracle on earth, absorbing carbon dioxide from the atmosphere, capturing external energy from the sun and storing it through photosynthesis, and in the process producing oxygen, thus making other life possible on the planet.

We look at the fundamentals of how solar energy and water are used by plants, as they have done for many millions of years. We then connect these processes to the substance and challenges of everyday life. At the risk of seeming repetitious we have tried to make topics complete, minimising cross referencing. We have not cited sources, but promise they exist. You will also meet key concepts in managing plants, for instance, water management, vegetation growth and ecosystem management.

Along with identifying the basic processes we also look at some important current issues. We believe our book should help you the reader to articulate all essential concepts, and to apply this knowledge to effectively understand both current and upcoming challenges.

David Smith & Duncan Rouch

Preface to the revised edition

I was brought in to revise this book following the sad death of Dr David Smith in 2017. We recall David Smith as an internationally renowned expert on agriculture and its relationship to the environment. In this revised version we bring an international context to the book. We have also tightened up and significantly restructure the text for improved clarity, while keeping the conversational sense of the original text, for the love and passion of David to plants and agriculture. We have also updated a number of figures and data.

To help you review what's in each chapter we added summaries of key information at the end of each chapter, called, 'Takeaway messages'. Finally, as also a reference book, we have added a list of topics, list of images and a glossary of key terms. We hope you enjoy this revised book.

Duncan Rouch, December 2018

Introduction

Rain and shine, water and sunlight, are two apparently very simple things that we recognise from our earliest childhood. They are the essence of our environment, determining our comfort, and underpinning all aspects of our very existence, as the basis of life on planet earth. Comment on these often opens conversations, for nice shine, 'It's a lovely day!' But we are less appreciative of some other aspects, 'I wish this rain would stop' and so on. We sing about it, 'Singing in the rain', 'Turn on the sun', and have rain dances. In the right combination we, 'Have a nice day'. In other rain and shine combinations we are uncomfortable and, if they are unexpected and we are not protected, people may even die, as many did in Europe in the extreme heat of the summers of 2003 and 2006. Also, of course, excessive cold causes many deaths from time to time. Plants respond too, to rain and shine, as thriving in the right combinations, or doing poorly, even dying, in others. Plants, too, can 'have a nice day', or die.

To gain a general view of the key factors that support growth of crop plants we first look at the origin of agriculture, about 10,000 years ago. Early people were roaming in ecosystems as they found them, as naturally occurring groups of other living things. How did they decide where was a good place as a central home area? Though the 'first' people continued foraging in a manner that could hardly be called a system, as hunting and gathering as earlier primates had. No doubt there was soon some system in the approach to maximise gain from effort. To begin with a recognition of the most likely places to find plants with more fruit, better fruit, and knowledge of the feeding and drinking habits of animals. As humans became worthy of the name, we suggest that managing plants and animals came together, and were mutually reinforcing.

Though we imagine this era to have been one of small numbers of people with a large area to hunt and gather, such things are relative. The variability of the effects of shine, and falling of rain would have meant alternations of plenty and scarcity, even famine. As a result, people may have learnt that a clearing in the forest allowed better growth of crops, under more shine, and crops survived better in dryer times if water from the local stream was used to water the plants.

Nevertheless, it is not hard to imagine a situation where, with the best will in the world, gathering would have quite a severe impact on an ecosystem. For instance, the needs for survival of the humans could be mandated to support breeding animals, just as in poor developing countries, as their staying alive takes precedence over setting aside grain as seed for the next crop.

Then there is the question of soil fertility, of the supply of essential nutrients for plant growth. Though some plants are more tolerant of lower levels, they can never completely do without essential elements. Soils derived from certain sorts of rocks, like the very mixed glacial deposits so common in Europe and North America and some recent basalts, benefit from a steady release of a supply of these elements. Large areas of the earth's surface, however, have parent material that releases nutrients very slowly, or has little to release. For instance, in southern Australia there were large areas of almost pure sand, low in all nutrients, and a very specialised group of slow growing plants had evolved to grow on them.

Changing the level of a nutrient would make possible the flourishing of some different plants, for example, using animal or human waste as an organic fertiliser. Also, a new plant arriving, perhaps carried by a bird from far away, or in the gut of a roaming animal, might have been even better suited to the local ecosystem, and over time come to dominate the area.

For a small hunter-gatherer population roaming an area with steady release of nutrients from the parent rocks, rainfall well spread through the year, enough shine to keep the soil warm in winter, and therefore no wild fluctuations in food supply, would seem to be a rosy scenario. On the other hand, if nutrients were released very slowly, and rainfall was variable and not well spread and, and for some reason there was a larger hunter-gatherer population seeking support, for instance, through tribal conflict, then every element of the ecosystem, including the humans, could be under stress. We can imagine there being some competition for hunting on the areas of better resources. So, we can argue cogently that hunter-gathering was a stage in evolution rather than a system to which we would want to return. At higher population numbers it could be very stressful.

In this story we see the emerging knowledge about the importance of rain, shine and fertilisation, to improve both the reliability and yields in producing food crops. To understand these factors it is vital to understand the underlying basic science, so, before we deal with the practical subjects, some comment is appropriate on the knowledge itself.

Each field of scientific knowledge can be imagined as a pyramid: a base of basic scientific knowledge, that includes a sort of language with an 'alphabet'. From this are derived a set of concepts, definitions and laws, Image 1.

The scientific pyramid is loosely connected to an inverted pyramid of common sense. This includes current notions about the practical use of technology in daily life and then, in a sense resting on top, the beliefs widely held by the community at large.

Unfortunately, common sense notions and beliefs can break away from the base, their roots, and have a continuing life of their own, growing branches of new 'ideas', that, if tested against the base, are easily shown to be misleading or even very wrong!

So, throughout this text it is our aim to relate current issues to the basic fundamentals, verifying the connections. It is then important to use this as a basis for evidence-based reasoning.

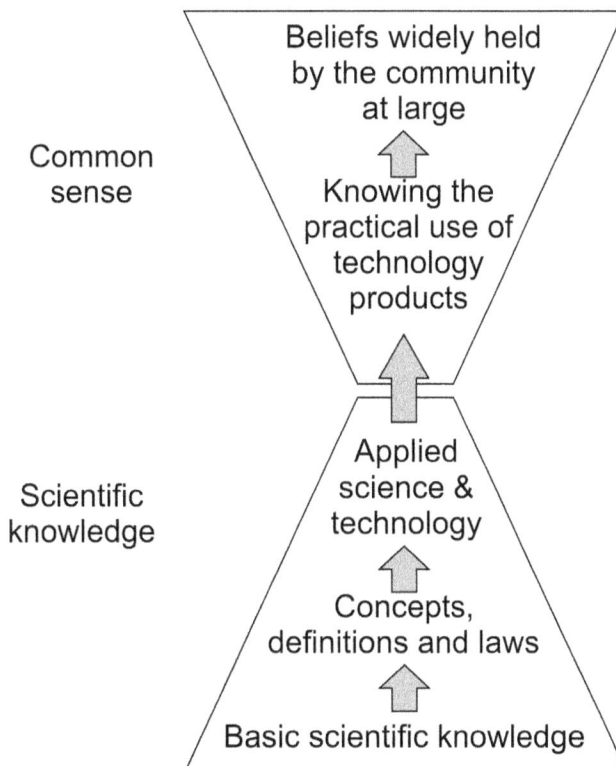

Image 1: The relationship of scientific knowledge to common sense.

The book is in two parts. Part I, headed What Makes Plants Grow, defines and discusses the processes of rain and shine falling on planet earth, and how these help to proliferate plants and associated life. Part II, headed How We Grow Plants, analyses the way humans manage the interactions of rain and shine with plants to fulfil human needs for food, as well as fibre, shelter and water.

Part 1

What Makes Plants Grow

1

Shine

Shine is the sun's rays. The Sun is a massive entity, brimming with energy, which radiates, to send huge amounts to Planet Earth, some of which can be absorbed, or reflected, by any object. The percentage of absorption and reflection naturally varies over a huge range, mostly to do with the nature of the surface of the object which receives the rays.

1.1 Solar energy

The sun is an enormous nuclear reactor with a surface temperature of about 5,500 °C. The shortwave radiation that it emits is the source of nearly all energy on earth, the exception being the chemical energy that holds minerals together in rocks. Coal and oil are simply stores of solar energy that were captured by plants, or by other organisms that depended on the plants, at various times since green plants emerged on the planet, roughly 300 million years ago. Wood is material from more recent capture of solar energy by living plants. Variations in the absorption and reflection of solar energy drive wind and rain. Solar energy radiation provides the energy for water to become water vapour, to form clouds and then rain.

If we capture some of this, then run it through a turbine we make hydroelectric power. That is, we use solar energy that has been used to elevate the water as clouds, then arrives on earth at a higher altitude, thus having potential energy to be released as it runs downhill. Also, winds, caused by differential heating of the earth's surface by the sun's energy, can drive a special design of turbine for wind power.

The amount of the sun's energy reaching any site on the earth varies from zero at night up to a maximum of about 900 watts per square metre when the sun's rays are at right angles to the surface and is not interrupted by dust or clouds. Roughly a third is reflected back into space at once, the rest warming up the land and water surfaces and clouds, which quite soon radiate about another third back into the atmosphere.

Green plants obtain energy from sunlight through their leaves, which contain interceptor cells that we call chloroplasts. These cells contain chlorophyll, which is a coloured compound that absorbs sunlight and gives leaves their green colour. Also, silicon cells, one type of which we put together to make what we call solar panels, are the industrial equivalent of these leaves.

1.2 Vital life processes

Photosynthesis

Plants utilise sunlight through a process called photosynthesis, which is the production of a carbohydrate (CHO), glucose, in the presence of sunlight. Though it is a complex process achieved in a number of steps, the overall equation is: $6CO_2 + 6H_2O + se \rightarrow C_6H_{12}O_6 + 6O_2$.

In words, the green plant uses solar energy (se), to break down the water (H_2O) to H and O, and combines these with CO_2 to form glucose ($C_6H_{12}O_6$). This simple sugar, glucose, is the building block for a variety of carbohydrates, including the major ones cellulose and starch, which are used to build plant structures, such as leaves, stems and roots. Glucose also supplies the energy for the metabolic processes that maintain life (see respiration below in this section). Most plants deal in carbon as multiples of C6, but a few specialised ones utilise C4 compounds. The oxygen collectively released by photosynthesis maintains the air at a suitable composition for breathing by animals, including humans. In capturing solar energy and storing it in a form that creatures can use, and providing oxygen for the needs of life, photosynthesis of green plants must be seen as an awesome process.

Most green parts of plants can photosynthesise but leaves do it best. The green colour is due to the presence of chlorophyll, the compound that actually absorbs solar energy and acts as a catalyst in the process, that is, it can be recycled. Leaves come in a variety of shapes and sizes but always have certain features such as specialized pores, called stomates, through their surfaces so carbon dioxide can diffuse in to the sites where there is chlorophyll. The water needed for the transaction is drawn into the plant through the roots, and up to the leaves by transpiration, section 2.4. Leaves are very often flat and thin and green, to serve as efficient interceptors of solar radiation and accumulators of solar energy. The structure of some plants and leaf shapes are sketched in Image 2a to d. One plant group, the legumes, including the clover and lucerne plants and the broad area bean and lupin crops, Images 3 and 4, have a bilateral deal with some bacteria, to access the vast nitrogen (N) store of the atmosphere, for enhanced production of protein. This has turned out to be a very vital process, which we will discuss later in more detail, section 4.2.

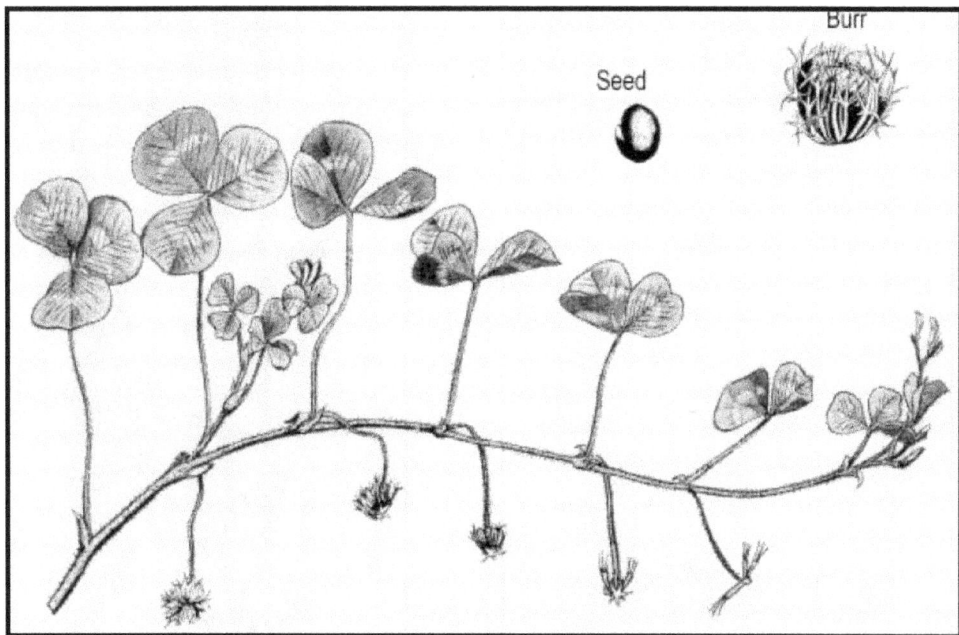

Image 2a: Subterranean clover.

Crop studies suggest the maximum efficiency of solar energy conversion to chemical energy at the sites of fixation in the leaf is about 10%, incorporated as chemical energy in the carbohydrates. Over time, hundreds of millions of years, the plants have evolved a huge variety of forms and structures and storage organs, this last often to do with their own long-term survival. We adapt these harvestable organs for our use, that is, seeds, which in the case of cereals we call grain, tubers in the case of potatoes, and fruit such as apples. Including this transaction, the greatest solar energy efficiencies recorded by crops are down to around 5%. We humans, very late but clever arrivals, have used these processes and organs as our food base, at first selecting, then deliberately improving, shape and size and taste. Inevitably these organs have also been attractive as a food source to animals, birds, insects and micro-organisms. We often do not wish to share with these, so devise ways of crop protection. This is, ideally, genetic, either physically or chemically repulsive, sometimes physical separation, as in netting, and sometimes chemical additives to kill the other creature or organism.

Image 2b: Lucerne.

Image 2c: Perennial rye grass.

Image 2d: Kikuyu grass.

Image 3: A very dense crop of the legume Faba beans, with author Dr David Smith.

Image 4: A dense crop of lupins.

Overall, due to limitations of temperature, water, or nutrients, however, the average efficiency of vegetation is very small, less than 1 %, but there is so much greenery that the impact of photosynthesis on our environment is enormous. Most crops easily achieve a growth rate of 200 kg/ha/day, that results in extracting the equivalent of all the carbon dioxide to a height of 70 m above it.

The inbuilt capacity of seedlings to emerge from the soil, make some chlorophyll, and immediately photosynthesise by plugging into the only energy source they need, the sun, is the primary miracle of our world, after the spark of life itself.

Photosynthesis is actually quite sensitive to the strength of the sun falling on the plant. We all know 'shade' as something blocking the sun from directly shining on us, out of the sun. We also know deep shade, as many layers of leaves between us and the sun, but few recognise its importance in plant growth. In winter, a horizontal leaf may cut out a third of the sun's energy, so under three layers it may be as dark as night. Lower leaves may not get any shine at all, but as these have to be 'kept' by the plant, these and their stems continue to respire. Then, these may be a net liability, in energy terms. This is an important concept for crop and pasture production as a 'deep' pasture sward may look good but not be making any gain in biomass, and so be quite unproductive. See Part II, especially Images of field crops (3, 4, 16, 21, 24) and pastures (7, 22). The same can be true of a forest. All the good work of the young top leaves may be undone by the lower parts of the tree, which become effectively parasitic. This has important implications for storage of carbon in considering greenhouse gas management and global warming.

The carbohydrate compounds of plants are thus an energy store, with the energy accessible at any later time by virtually reversing the process. If it is millions of years, we call the plant remains fossil fuels. As noted above, some of the reversal happens immediately, as all of the time all living cells, plant and animal, stay alive by producing energy in respiration. The gas carbon dioxide is, of course, produced by all other oxidations of hydrocarbons, including car and truck engines, coal and gas burning power stations, forest fires and domestic wood fires and stoves. Here is a feedback loop, in that the rate of photosynthesis can be speeded up by higher levels of carbon dioxide in the atmosphere. Interestingly, some producers of flowers and fruit grown in greenhouses use carbon dioxide enrichment to accelerate growth.

Respiration

Carbon dioxide used to produce carbohydrates through photosynthesis is ultimately replaced by respiration, by the crop itself or the organisms that eat or decompose it. Respiration is the oxidation of carbohydrates, like glucose, which provides the release of chemical energy to maintain the metabolic processes of living things. As an overall chemical reaction, it is the reverse of photosynthesis, that is: $C_6H_{12}O_6 + 6O_2 -> 6H_2O + 6CO_2 + ce.$

Put into words, the glucose $(C_6H_{12}O_6)$ is oxidised, broken down to its component molecules, water and carbon dioxide, and the stored chemical energy (ce) is released. Respiration is even more widespread than photosynthesis because the cells of all living organisms, including humans, rely on this process to supply energy to maintain life. Plants obtain glucose from photosynthesis and oxygen by diffusion from air, especially through the stomates of leaves. Animals, like us, obtain glucose from the food they eat, either directly or indirectly originating from the photosynthesis of plants, and get oxygen by breathing air into the lungs, there combining the oxygen absorbed into haemoglobin and distributing it throughout the body in the blood stream. Plants have no active system for distributing oxygen though some specialised plants such as rice have evolved cells in their root systems that facilitate internal diffusion of oxygen which allows them to grow in waterlogged soil.

In plants, energy from respiration is used to maintain the cells in good working order, able to carry out a range of processes, for instance, nutrient uptake. This enables growth, that is, the construction of new chemical compounds from which more leaves, stems, and roots are formed. The amount of energy needed to maintain a plant depends upon its size and also temperature because chemical reactions run faster as temperature increases. Clearly, plants are able to grow only when photosynthesis exceeds respiration. That is, when more CHO is made than is used. Also, as plants grow larger, the amount of respiration needed to maintain the plant increases. Then, unless the plant can maintain actively photosynthetic leaves that intercept light, growth will slow and may cease. Many crop plants, in the later stages, divert all products to seed, so without new leaves to maintain the increasing demand for respiration, they die, a process we call senescence. If they do this every year we call them annuals. Understanding the relative functions of these and perennial plants is important, and are discussed later, sections 4.2 and 6.14.

The huge importance of the element phosphorus (P) in all of the energy transactions should be noted. It helps forming adenosine di- and tri- phosphates,

ADP and ATP for short, and so is involved in stages of the equations for photosynthesis and respiration. Though some plants have adapted to operating at a slightly lower level of P, a low content of P in soils generally spells a low level of plant activity in the energy system. A search by some bio-technicians for plants that do not need fertilisers such as P is rather like seeking a way to defy gravity!

1.3 Organic compounds

The CO_2, (O-C-O) as the raw material of photosynthesis and for the product of all carbohydrate and hydrocarbon oxidations, is like H_2O (H-O-H), just about everywhere there in life, as a vital part of the life system. The three elements C and H and O form a vast array of organic compounds in nature. Humans have manufactured many more, which we call synthesising, and these compounds synthetic. A few examples of the compounds occurring in nature will illustrate their importance. If H-O-H has a CH_2 added we get CH_3-O-H methyl alcohol, also known as methylated spirits, which is not an odourless, tasteless compound like H-O-H, but a fairly nasty one. Add another CH_2 and we get C_2H_5-O-H ethyl alcohol, the compound many humans consume every day, which we simply call 'alcohol' for short. Or, bring in a COOH and we get CH_3-COOH, acetic acid, which we call vinegar. Add another CH_2 and we get C_2H_5-COOH, a larger organic acid.

Or, build on CH_2s (with no O) and we get a string of hydrocarbons (HCs), like our fuels and oils. Starting with a low C number we get gases, with more Cs high octane liquids and, as more then, longer chains of Cs, as oils and greases. The lesser carbon HCs all burn, as we know, oxidising, to give CO_2 and H_2O.

Or, instead of forming a chain, the Cs can play ring-a-ring-a-rosy, with six of them called a benzene ring, C_6H_6. Various things can replace one of the Hs, for example chlorines (Cl) at Cs 2 and 4 and an acetic acid at another C and we get 2,4 D, that is, 2,4, dichloro phenoxy acetic acid. Add three Cls and we get 2,4,5 T, that is, 2,4,5, trichloro phenoxy acetic acid. The phenoxy acetic acids are a group of compounds naturally occurring in plants known as auxins, which rather like catalysts, are very important in plant growth. Of course, we all know 2,4 D and 2,4,5 T as herbicides with a bad name, and prohibited in many countries. Nevertheless, it is important to understand the close connection of these compounds to natural compounds. This similarity to a fair extent explains how we can use them as herbicides by raising the levels in a way that accelerates the metabolism and finally 'messes up' or confuses the weed plant, so that it dies.

1.4 Balance of atmospheric gases

The importance of plants evolving and then proliferating on earth can be understood when we realise that before their existence there was no free oxygen. Back then any oxygen present was immediately taken up by oxidation in fiery processes. Once plants produced enough of it for their own respiration, these then prospered enough to make a surplus which accumulated in the earth's atmosphere. Then organisms without their own energy capture systems could become possible. One group is animals, including humans, who are in that sense wholly plant dependent. In turn, such animals 'counter' the rise in oxygen by using it up to make carbon dioxide again. So, we should also note the flux of carbon through its various combinations in compounds through the air, the plants, the creatures, the soil. What we do, in managing our activities, can profoundly affect the levels and amounts in any one stage, and perhaps lead to global warming through high levels in the atmosphere of carbon dioxide and methane, which are both carbon compounds. We will deal with this question later in some detail, section 5.4. Incidentally, everything mentioned above, from water on, is a chemical.

1.5 The earth and the sun

Since early scientists realised the earth was round, and the Polish scientist Copernicus established that the sun, rather than the planet earth, is the 'centre' of the universe and planetary system, we have learned a lot more about the interactions of the earth and the sun. We understand night and day and its changes. We recognise that climate varies from place to place over the earth and can be manifest differently from time to time at any one place: these latter differences we call seasons. We are aware of interruptions to shine by clouds and shade cast by other things on earth. We know about air movement, wind, and understand that these are largely a consequence of shine and differential temperatures across the earth's surface.

We realise that, as well as spinning, the earth rotates about its axis, so it seems to us that the sun moves north and south through the year, thus changing its daily east-west arc from directly overhead at the Tropic of Capricorn, the summer solstice and season in Australia and winter solstice and season in North America and Europe, to directly overhead at the Tropic of Cancer with reverse seasons. So, 21-22 December, longest day, and 21-22 June, shortest day, are important dates, and are highly reliable in terms of changes in shine. This broadly brings about the seasons, summer when the sun in most overhead, and winter when it is least so. Storage of heat by the earth, and even more by water, means there is a lag, so the hottest weather in the southern hemisphere is early February, six weeks after the longest day, and the coldest is early August, about six weeks after the shortest day. In the northern hemisphere the reverse is the case. One result is that equal length of day and night at the equator occurs twice a year and nearer to the poles there are increasingly longer summer days and shorter winter days. There is more land in the northern than the southern hemisphere at higher latitudes, and so the very long summer days in the northern hemisphere are profoundly important for total plant growth and crop yield. For instance, corn is grown right up to Hudson's Bay in Canada, and barley ripened in Iceland at 50 degrees north. In the southern hemisphere there is little land at similar latitudes, the most intensive cropping being near Hobart in Tasmania, Australia, at about 45 degrees south.

We also recognise that there is a large difference in total solar energy received on a clear summer day and on a clear winter day, when the sun is lower in the sky at its zenith at noon. This different amount of energy available for capture by plants between summer and winter greatly affects plant growth. The seasonal variation in day length has a lot to do with plants flowering, in particular by the lengthening days of spring. Many studies have shown that variation in day length has prime importance in the adaptation of crops to specific locations and also to crop management by humans.

1.6 Photoperiodism

The control of plant development by lengths of day and night is called Photoperiodism. During evolution plants must have 'found' long ago that day length is a more reliable indicator of seasons than temperature or rainfall. This is confirmed in our time by the fact that weather reports can predict sunrise and sunset for any future date to the minute, whereas at most latitudes a given date can have very different temperatures year to year. Rainfall is even less predictable. Nevertheless, the exact date of flowering varies a little from year to year in some cases because other factors, such as temperature, interact to some extent. One interesting example is subterranean clover, a very important legume pasture plant in southern Australia, which must accumulate a certain amount of 'cold days' to change from growing leaves to flowering.

We call plants that have flowering triggered by days lengthening (Spring) long-day plants and plants that flower as days shorten (Autumn) short-day plants, reflecting the environments in which these plants have evolved and adapted. Those from mid-latitudes tend to be long-day plants while short-day plants mostly originated in the tropics. Actually, we now know that plants use the length of night to determine the length of day, called the photoperiod, using a pigment system located in leaves. This was clarified when it was observed that a light break of just a few minutes during a long night is just as effective in promoting flowering of long day plants as lengthening the day. Application of this knowledge is used in horticulture, when night time light breaks are used to hasten or synchronize flowering. The reason for this is to have batches of plants that flower together. In this way plants can be brought to flower ready for market on a special day such as Mothers' Day.

Flowering isn't just a 'pretty' thing and there are many plants that do not have colourful or attractive flowers. Rather, it is an important stage marking the beginning of reproduction, a process that involves fertilisation and, if conditions are right, ends with ripe seed and so ensures continuance of the species even if the parent plant dies. Sometimes, especially for annuals, it may be a great mechanism to assist spread to new sites. For these annual plants, it is critical that flowering occurs when temperature is suitable and there is adequate moisture for growth to fill and ripen the seed. Photoperiodic responses tune species to particular sites as seen in the small differences in the many land races of crops in traditional agricultural systems.

In the search for greater productivity, plant breeding may seek cultivars without a photoperiodic response, to broaden their geographical adaptation and also to allow out-of-season cropping. For example, the new rice-wheat production systems have no photoperiodic responses, and so can give greatly increased food production across the Indo-Gangetic Plain of the Indian sub-continent.

1.7 Influence of topography and clouds

Differences in solar energy falling on a plant can be confounded by slope. In winter, in the southern hemisphere, the difference between a slope of 20 degrees north-facing and a 20 degree south-facing slope may be nearly as much as the difference between summer and winter on a flat plain. On the other hand, slope has little effect near the Equator, for instance on the steep land in Rwanda and Brazil, shown in Images 5 and 6, where the sun is more nearly overhead throughout the year.

It is interesting to look at various pictures of landscapes and consider when and where slope is a factor: for instance, comparing the above two in Rwanda and Brazil, and Image 7, in the Adelaide Hills, South Australia.

Image 5: Farming on steep slopes near the continental divide in Rwanda.

Image 6: Cropping on steep land, south of Rio de Janeiro, Brazil.

Image 7: Grazing land with river red gums in the Adelaide Hills, South Australia.

Then there is the cloud effect. As we know so well, in winter a warm sunny spot can quickly become cold and inhospitable when a cloud passes between us and the sun. Overlay this difference in shine with lower temperatures and we can understand why plants grow more slowly in winter, especially in high latitudes. Closely related to this, though we may not be very aware of it, is the total use of water by plants. This is called evapo-transpiration (ET), an important item in the quest for efficient use of water by crops, which will be discussed fully in section 2.4 and 2.5.

Take away messages

1. The two plant processes, photosynthesis and respiration, are of critical importance to sustaining life on planet earth.

2. The three elements C and H and O form a vast array of organic compounds in nature.

3. The variation in shine and temperature across seasons has a substantial effect on the growth and flowering of plants.

4. The control of plant development by lengths of day and night is called Photoperiodism.

2

Rain

In considering rain we need to see it in the context of the massive amount of water on the planet.

2.1 Water on earth

We all know from our early childhood that rain is water, that the sea is water! Almost all of the water on earth is in the sea. About 2.5% of the water on earth is 'fresh', that is, not in salty seas and a bit more than 10% of this, that is, a little over 0.25% of all water is available for management by humans. The rest is 'locked up' in icecaps, glaciers and snow fields. It tends to pervade all sorts of situations, even be locked away for vast ages in these icy situations, but also in the earth between strata in underground streams and 'lakes' known by the term water-table or aquifer. Some are deep, out of reach of humans: some get squeezed out as springs; some just linger near the surface, perhaps even in the root zone of plants, assisting the growth of some, impeding the growth of others. Some have little impurity, and some are very high in minerals. Some bodies of water are cold, and some are very hot, as in places like Iceland and New Zealand, where energy is extracted from it and electricity generated. There is increasing interest in the deep hot water as a source of thermal energy and electricity generation.

Water also 'fits' into the spaces between soil particles, with minerals from the soil particles going into solution. The relationship of the soil particles to water, and the capacity of the soil to store water is of enormous importance in the growth of plants and management of crops. So, much attention will be given to this later, section 2.4.

2.2 What is water?

Textbooks describe water as a colourless, odourless, tasteless liquid, a harmless, inert substance, the universal fluid, which is the base for animal blood and plant sap.

Water is simple, H_2O, or H-O-H, an oxygen atom linked to two hydrogen (H) atoms. It occupies a strategic place in the structure of chemicals. Replace one of the H atoms and the O with a chlorine atom (Cl) and we have hydrochloric acid (HCl), a highly corrosive substance, though which is present in our stomachs, and filling an important role in digestion. That it does not eat away our gut is a tribute to our biological design! Replace one of the H atoms with a sodium atom (Na) and we have sodium hydroxide (NaOH), another very corrosive substance, commonly called caustic soda. We could go on. There is a raft of organic and inorganic chemicals, for the stuff of life, and sometimes death, on earth.

Water is the liquid form of H_2O. At about 100 °C, the exact temperature depending on atmospheric pressure, H_2O becomes a gas, that is, water vapour, which is visible as steam when it is condensing in quantity in cold air. At about 0 °C H_2O becomes a solid, which we call ice. We humans, and the plants and animals that have evolved from early living material on earth, fit fairly nicely into this range 0 to 100 °C, not surprisingly, as water is the base fluid and all things have evolved using it.

Water is also a heat store and carrier. It takes heat to raise its temperature, and if we insulate it properly we can store this heat in it. Then there is the change of state, from solid to liquid, and liquid to gas. It takes quite a large energy input to make these changes, so warmer water has energy stored in it, and steam much more, and there are opportunities for us to utilise this, for instance, in thermal power. So, the energy dynamics of water are very important in energy management, a large subject in itself. One interesting application in farming is the dairy farmer who needs to cool the cows' milk so it will keep better, and he or she would like the water from the tank warmed for washing the hands of the workers and the udders of the cows. Then, a heat exchanger is used to effectively transfer the heat from the milk to the water. Heat pumps involving this aspect of water have widening application.

Water brings life, as essential to it, but it can also bring death, not only through drowning but also because it so easily carries toxins and harmful micro-organisms. It has been suggested that over the whole planet 80% of human infections and illnesses are water borne. Paradoxically, life brings water; living things actually manufacture it, too, as a product of the process of respiration.

Water carries things. Water is the carrier of needs, the life fluid. Living things have their own needs for substances that may be contained in water. They have evolved to use "impure" water. Plants use soil water with substances dissolved in it. Animals too need minerals, and at times farmers actually add minerals to the drinking water of live-stock. Humans buy 'mineral water'. However, there is a fine line between enough and too much, in levels of sodium chloride (common salt) for instance, and in section 2.8 we will discuss the common problem of too much salt in water in southern Australia.

Water too dissolves material. It is then called a solution. When cooled it may not be able to hold all of the dissolved material, as some may be precipitated out. However, even near freezing point, water can still contain some dissolved material, though this will reduce the freezing point. In this way the water of a fresh lake may freeze but the nearby sea not do so. Water also has the peculiarity of expanding when it freezes, so ice floats.

Though water containing impurities will continue to contain these in its solid form, ice, the change from liquid to gas is such a dramatic physical rearrangement that water vapour is a clean product, which when condensed gives very pure water. This cleaning is a very important function on planet earth, in essence a renewing of our water.

2.3 A lot of water – so why is it scarce?

Yes, there is plenty of water on earth, but so often it is the wrong quality at the wrong place or wrong amount at the wrong time. Though water is colourless, odourless, tasteless and harmless when pure, the problem is that, being the universal fluid, it is often has things in solution and in suspension, very often because we have either accidentally or deliberately loaded it up with unwanted or dangerous substances. We use it as a cheap and convenient carrier to wash away unwanted things, but do not often accept responsibility to restore to clean it and return it to nature in appropriate condition. The natural hydrologic cycle has its own restorative processes, so is naturally sustainable, through evaporation and condensation, through natural filter beds, through slowing down flow and depositing suspended material.

2.4 Soil water

Plants require water for growth and survival but respond only indirectly to rainfall. To plants, the important feature of water supply is what is available to them in the root zone of the soil because it is the storage of water in the soils that maintains plants from one fall of rain to the next. Rain that falls adds water to the soil while drainage and evapotranspiration remove it. We define evapotranspiration by the following equation.

Evapotranspiration (ET) = transpiration + evaporation

Transpiration is the process of movement and transfer of water from plants, from pores in the leaves, to the atmosphere while evaporation is the transfer of water from soil to the atmosphere. Image 8 introduces some of the concepts of rainfall patterns, accumulation of water in the soil, and this then being used up by plants.

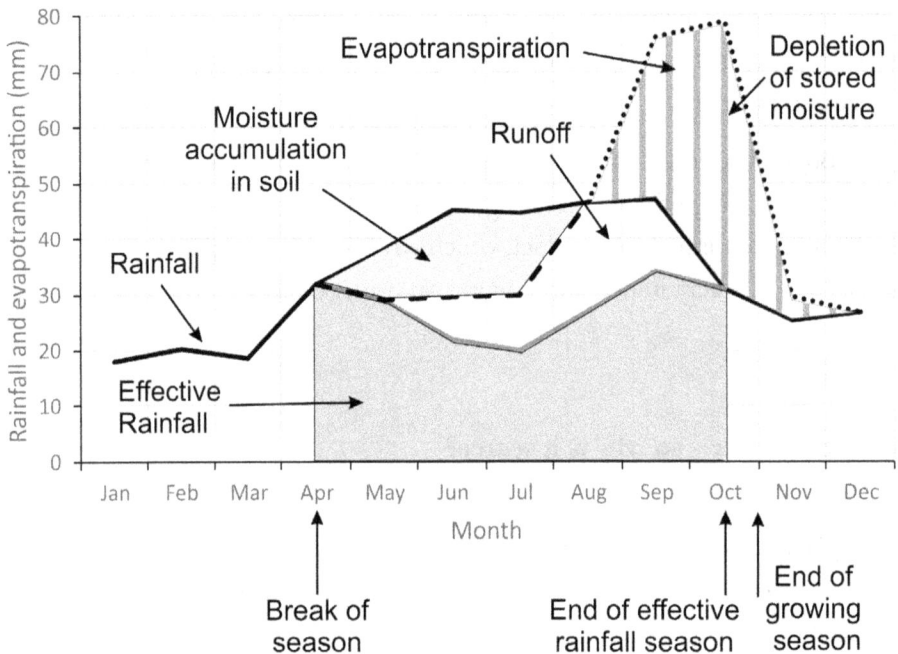

Image 8: Moisture relationships at Roseworthy, South Australia, showing the growing season and the importance of soil water. Effective rainfall is the part of rainfall used by plants. The growing season is from mid-April to the end of October.

The amount of water that soil can hold between its particles in a freely draining state, the upper limit, and the lowest point at which plants can extract water depends largely on soil texture, with organic matter the main modifying factor. Sandy soils hold little water while clays hold most. Image 9 presents a notional soil profile, as a vertical section. Water would tend to soak in and accumulate in the B horizon. The water holding capacity per unit of volume, together with soil depth and the rooting capacity of individual species of plant, determine how long plants can resist drought and also tells us how much and how frequently to irrigate for high productivity.

Soils will become waterlogged, that is, be so filled with water between the pores that there is little air in the root zone if an impermeable layer such as clay or rock restricts or prevents drainage. Waterlogging is a serious problem because, with few exceptions, plants, including their root systems, cannot live without oxygen and those that can do not grow actively under those conditions. Waterlogging is more common in high rainfall areas but can also pose important problems in semi-arid areas where, though the total annual rainfall may be small, it may fall in a few big rains, or be concentrated in one part of the year. Waterlogging may also connect the root zone to deep subsoil moisture, and if there is salt may mobilize it into the root zone, a further problem.

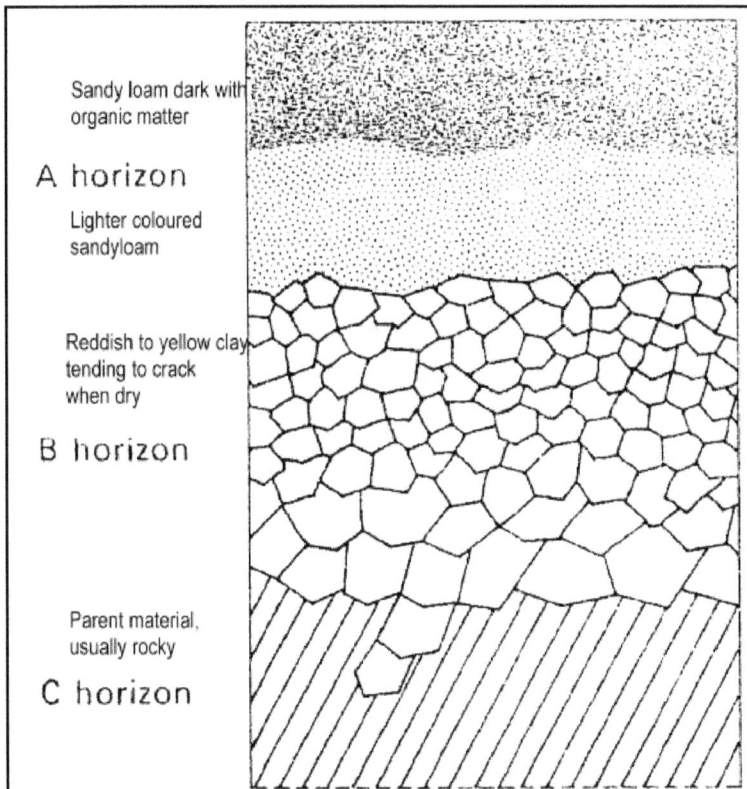

Image 9: Common arrangement of horizons in a soil profile. This is an example of a duplex soil, which has a sudden change from a loamy A horizon to a clay-filled B horizon.

2.5 Movement of water through soil and plants

Water movement through the system is a vital part of the processes of life. Some is used as a physical entity: acting as a solvent and in energy management. Some is used as a chemical entity, as a compound of hydrogen and oxygen, to supply these elements for plant processes.

In transpiration, water from soil carries nutrients in solution, flows through the roots and the stem to the leaves, which we call it sap. Leaves lose some water, most of the time to maintain them at an appropriate temperature by evaporating water from their surface. The energy for change of the water from liquid to gas is taken from the leaves.

Plants have evolved as entities that have a process of fixing solar energy, by combining water and carbon dioxide to form glucose. The stomates, pores in the leaf surface, must open to allow inward diffusion of carbon dioxide and to allow water to evaporate and move to the drier atmosphere outside.

So, a useful way to view plant growth is to see it as an exchange with the environment of water for carbon in the form of carbon dioxide, though the plant must do a lot of work to make this happen. That is, less than 1% of the water absorbed by plants from the soil enters the process of photosynthesis and so is retained by the plant, while all the rest is in the transpiration flow. That small percentage is of enormous significance for life on earth.

Plants have developed a variety of adaptations to reduce water flow through transpiration: some can close their stomates, some have small leaves, some orient leaves to reduce the sun's energy impact, so there is a lower need for evaporation. Others have some clever mechanisms: an outstanding example is succulent plants common in desert communities which have their stomates closed in the presence of sunlight, but stomates open at night when evaporation is small and absorb carbon dioxide into organic acids. Plants with these various adaptations use much less water per unit of carbon fixed than other plants. The most important commercial plant of this type is the pineapple.

As well as the above uses of water in individual plants, water is also lost from all masses of plants, including crops and pastures, by evaporation directly from wet foliage and especially from wet soil. The term evapotranspiration collectively describes these two losses of water: transpiration by flow up the plant, use in photosynthesis and respiration, and evaporation from the soil surface.

Together solar radiation and air movement, wind, help maintain a humidity gradient from evaporating surfaces of plants and soil. In this way the loss of water to the atmosphere helps cool these surfaces.

The amount of water available to the plant in its root zone obviously affects transpiration, but plants can maintain transpiration at much the same rate even when the surface of the soil is dry as long as the soil below has moisture available. Evaporation from the soil itself depends upon the wetness of the soil surface, decreasing rapidly a day or so after rainfall. The balance between the size of the two components depends upon the leaf mass and the pattern of rainfall. Dense crops have high transpiration and also shade the soil, thereby reducing soil evaporation. Water will evaporate from the soil surface if the vegetation is sparse, including young crops, being especially rapid after rainfall, so the frequency of rainfall, rather than the total amount, determines the quantity lost by evaporation. A small number of large rainfall events will evaporate less and add more water to the root zone so that transpiring crops can maintain growth.

Water comes from clouds, as rain, hail or snow and the word precipitation is used to include all of these forms. Clouds form when water vapour in the atmosphere condenses into water droplets, still widely dispersed, but visible. And where does this water vapour in the sky come from? Some comes from evaporation of what we call free water surfaces on earth, liquid H_2O turning into gas H_2O, like steam, with the energy needed to make the change coming from the shine, the sun's heat. Evaporation doesn't mean destruction, only change of state. Other water vapour comes from plant transpiration and animal perspiration and, of course, some new water is made in the chemical process of respiration. That is, oxidation of carbohydrates to release energy, generates carbon dioxide and water.

At an early age we learn about rain being wet, and some wetter than others! We soon understand mist, drizzle, showers large and small, soaking rain. Gathered together, these form the rainfall pattern of the season, and largely determine when conditions are right for plant growth. That is, when enough of this has reached the soil and soaked in so there is enough moisture actually in the soil for plants to extract it. This is marked on Image 8 as the break in the season.

Plants require a continuous flow of water for growth and survival from one fall of rain to the next. Then, it is important that water is stored in that part of the soil which coincides with the root zone of the plant. Rainfall adds water to the soil while drainage and evapotranspiration remove it. There is a point at which plants wilt even if there is still some moisture left, but it is so tightly held, that plants cannot extract it: this is called the Wilting Point and expressed as a percentage by weight of the soil. Moisture falling on the soil does not always soak in, so especially when a lot of rain falls, some may run off into gutters, and even when it enters the soil some may drain through, as the soil can only hold a given amount. The percentage held under free drainage conditions is called the Field Capacity. So, a lot of rain does not necessarily mean there is going to be a lot of water available for plant growth.

The amount of water that soil can hold between the Field Capacity and Wilting Point, depends very largely on soil texture. Sandy soils hold little water while clays hold most. So, soil composition together with soil depth and the rooting capacity of individual plant species determines how long plants can resist drought. If supplementary watering, for instance irrigation, is available, these parameters are a guide to how much and how frequently to add water for optimal productivity.

The sun, as shine, plays a part in determining whether the rain that falls results in water being available for plant growth. It provides the energy for evaporation from a free water surface, so whether the water simply dries up from the soil surface while slowly soaking in. This drying up, evaporation, is increased by wind, when molecules of water breaking free from the surface are swept away. There is also variation according to the nature of the soil. For instance, whether a lot runs off or soaks in, and whether there is a layer of mulch on the surface to reduce evaporation.

2.6 Seasonal availability of water for plants

Life on earth has evolved to suit the rainfall and temperature regimes that the sun produces. The physical and chemical properties of water are of critical importance here because the metabolism of all living creatures operates in an aqueous, water based, phase. Too hot and water boils, too cold and it freezes. In practice, large parts of the world experience suitable temperature, within the range 0 to 40 °C, for a sufficient time each year to support growth of adapted species.

A growing season is defined by the time through the year when there has been moisture in the soil available to the plants, and temperature is not adverse for growth.

It is easy to identify the Growing Season in Image 8. We can easily measure free water surface loss using an evaporimeter, and it is considered that loss of water from the soil and plant system is strongly correlated, that is, one third of it. In the State of Victoria, Australia, Image 10, when rainfall is greater than 1/3 evaporation we can consider conditions suitable for plant growth. So, near Mildura in the north west region the area has a 5 months long Growing Season (May to September), while land near the southern coast at Portland has a longer 10 months Growing Season. Temperature may moderate growth at times when plenty of moisture in available, so it is useful to assess winter temperatures, included in Image 10. Temperatures below 7 °C inhibit growth. In a place like Iceland such conditions prevail for six months. The tropical Northern Territory of Australia, with summer rain and winter dry, has a Growing Season over November to March.

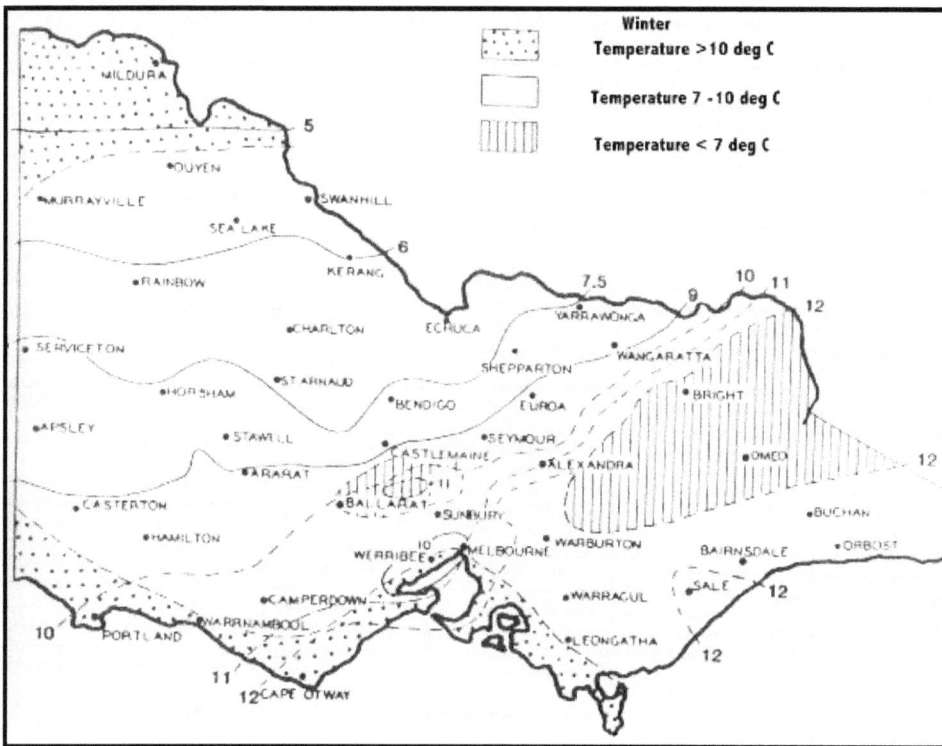

Image 10: Variation in growing seasons across the State of Victoria, Australia. Growing seasons are shown in months, and temperatures for the winter months are overlaid.

The agriculture practised in any area reflects these seasonal factors, though technology can moderate them, for instance as in heated green houses. It follows that each place will have a growing season determined by that year's rain and evaporation, which are closely linked to shine periods. Two key points in the annual cycle are; when soil moisture becomes available, called the break in the season, and when the end of the effective rainfall season occurs, which causes plant growth to cease due to exhaustion of water in the system, Image 8.

Moisture being available in the soil is a product of rain falling and soaking in, though not running off or seeping deep, nor being evaporated by shine and wind together. So, there is a good deal of emphasis on rainfall measurement, and though we work to averages and generalisations, each place in each year will have a specific growing season. A mismatch with plant growth on the dry side is labelled a drought, though the issue is very complex and will be discussed fully later, section 3.4.

Rainfall often varies a lot from year to year. If there is a deficit, it may be made up by irrigation where water is available: where there is surplus, it will probably mean increased run-off for dams. Also, even within what can be considered the same type of climatic zone, there can be big differences in monthly averages of rainfall, see Table 1. Consequently, it can be important to have local data on rainfall, rather than relying on regional averages.

Table 1: Average monthly rainfall (mm) in Mediterranean-type climate regions. These are classified in two main types, hot summers and warm summers.

Seasons		Summer			Autumn			Winter			Spring			Total
Months	Nth	J	J	A	S	O	N	D	J	F	M	A	M	
	Sth	D	J	F	M	A	M	J	J	A	S	O	N	
Hot Summers														
Valencia, Spain	Nth	22	8	20	70	77	47	48	37	36	33	38	39	475
Los Angeles, USA	Nth	2.3	0.3	1	6.1	17	26	59	79	97	62	23	6.6	379
Perth, Australia	Sth	13	9.5	13	19	44	118	177	170	134	81	52	22	853
Warm Summers														
Porto, Portugal	Nth	46	18	27	71	138	158	195	158	140	90	116	98	1,255
San Francisco, USA	Nth	3.3	1	2.3	7.1	30	84	81	120	105	86	32	14	566
Cape Town, South Africa	Sth	17	15	17	20	41	69	93	82	77	40	30	14	515

Northern (N), Southern (S)

So, to assess the effect of variation in annual rainfall we look at a local example, Image 11. This shows yearly mean averages of rainfall, using data over 106 years, for the Nagambie town located in the central area of Victoria State, Australia. This town is located just south of the northern irrigation area, so agriculture there relies on rain, so we call it rain-fed farming. To get a picture of short-term and long-term trends we use running means, Image 11. Because water is normally stored in the soil and plants may access quite deep water, a 'smoothed' running mean, for instance of a 5 year period, can give a more effective picture of the impact of annual rain. In practice the running mean period would correspond to the type of soil and shape of landscape. That is, a 5 years running mean would be useful for an area with rich soil that has a high water holding capacity, along with landscape that allows local collection of rain water to refill the local water table. In contrast, for an area with sandy soil and low water holding capacity a shorter running mean period might be more useful.

Looking at the long-term cycle, by the 20 years running mean, Image 11, we can see a rise in rainfall over time to about 1973, followed by a distinctly falling trend. This later decrease may be due to climate warming. Clearly local irrigation plans will include attention to both short- and long-term trends, including to assess what plants we can expect to grow at that location.

Many tropical environments have rainfall and temperature such that plants can grow all the year around, but that is less common at higher latitudes where cold as well as water-shortage restricts plant growth. In some regions, for example, the summer-rainfall continental climates of North America and Europe, the focus is on temperature because plant growth becomes possible with the thaw in spring and can often be maintained until it gets cold again in autumn. In semi-arid zones, where temperature is rarely a constraint, water supply, the balance between rainfall and evaporative demand, becomes the concern. Higher altitude, with its lower temperatures, imposes an overlay on plant growth patterns.

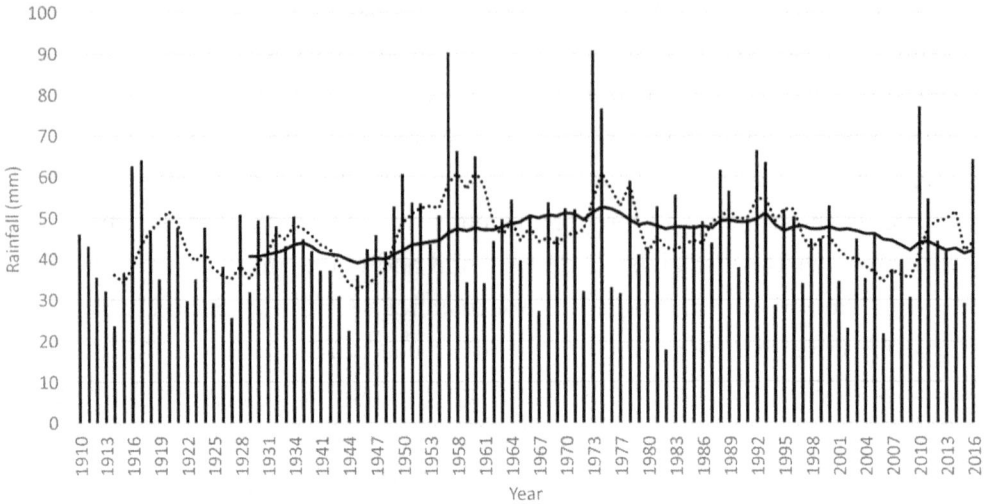

Image 11: Rainfall at Nagambie, central Victoria, Australia, 1910 to 2016. Legend: vertical bars, yearly mean average per month; dotted line, 5 year moving average; solid line, 20 year moving average.

Though the growing period may vary a little in length from year to year, provided the months are consecutive, there is good chance to devise and manage successful production of crops. This is the case in southern Australia where winter crops and pastures of annual species are grown over five to seven months of the winter-spring period. In semi-arid environments of the subtropics, for example, East Africa, there are two distinct growing seasons, each of three months, separated by a dry period of the same length so it is more difficult to grow productive crops.

2.7 Water and earth forms

Water plays a huge part in the constant reshaping of the earth's landscapes. In cold climates it may fill cracks in rocks and split them when it expands on freezing. Mostly its actions are due to that it, and wind, when moving over the earth's surface, are capable of carrying things. In this way it can profoundly influence the nature of the earth's surface. The size of particles carried is determined by the velocity of flow: mass varies as the square of water velocity. So there is the never-ending movement of small particles of the earth's surface carried by wind and water, often aided by gravity, and which we call erosion. This can be aggravated by human activity: Image 12 shows serious erosion after cropping near Mexico City, and Image 13 erosion near Benalla, Victoria, Australia. The removal of material from one place and massive deposition in another may even change the balance of whole continents, slowly sinking some parts lower in the oceans while other parts rise.

Image 12: Severe erosion following peasant cropping, Mexico.

Image 13: Water erosion of crop land near Benalla, Victoria, Australia, 1960.

The waters of oceans, arriving at a shore as waves, sometimes immense in power, smash the edges of continents, to change their shape. In our life time this is usually only a little, but over thousands of years, an immense amount. One huge question is to what extent global warming may increase this.

If we could see a time-lapse film of the changes of the earth over very long periods, millions of years, rather than the small span of 70 years or so of human life, we would see an incessant turmoil. Upheavals, which we call volcanos and earthquakes, along with pushing up mountains; breakdown of rocks by heat changes and gravity; flooding rains washing debris down the mountains; rivers carrying silt and sand, changing course over many kilometres; constant movement of particles in wind and water, sorting, building flood plains and deltas and reshaping the uplands; winds moving vast amounts of sand, not only along coastlines, sometimes forming sandy deserts. Sometimes, even in quite good rainfall areas, there are sand plains that, because of low fertility, carry sparse heath vegetation.

In recent times erosion has become more a loaded, dirty, word, than a recognised part of the cycling of materials on the planet. To some extent this has been brought about by what was judged excessive erosion caused by human activity, but also by lack of appreciation of its role in landscape formation.

The cycle of erosion

As flow of water slows it will deposit material: first stones, then gravel, then sand, then silt, then clay. So, water is a primary and natural force in shaping the surface of the earth and forming new areas of soil. A flood is a great source of new material, and ancient civilisations, for instance by the Nile river in Egypt, depended on deposits of these for renewal of fertility. Now, the great food bowls of the earth rest on flood plains: the Yangtse, Ganges, Tigris, Euphrates, Rhone and Mississippi, to name a few. Flooding remains a fundamental land forming process, and only becomes a disaster when inconvenient to humans, especially for those who have located their cities and farm homes on flood plains.

2.8 The hydrologic cycle

The water cycle is profoundly important, with evaporation, clouds forming over the sea, rain falling on land, water gathering as streams, then larger streams and rivers flowing back into the sea. When the rain falls on the land, some soaks into the soil, depending on the intensity of the rain, the slope of the land, the nature of the earth's surface, including plants, debris, which we call mulch, and the composition of the surface soil, in terms of sand, silt or clay. If the rain falls gently, spread over some time, an enormous amount of water may be taken into the landscape, with little run off. With larger falls most may run off, especially where there is little vegetative cover.

The water that runs off gathers into small streams, then bigger streams and, on most continents and locations, mighty rivers. The continent of Australia lacks really mighty rivers and is often described as the driest inhabited continent on earth. To illustrate, the Yangtze, arising in the north eastern Himalayas, is 75 times larger than Australia's largest river, the Murray! Some of the water that soaks into the ground goes down deep enough to join with other water, to form what we call ground water in aquifers. This may take thousands of years to emerge from the landscape at other locations, as what we call springs. The full cycle to the sea may take a very long time, or, in some cases, never be completed. The ground water may be intercepted by humans with water pumps and lifted back on to the surface for re-use. Springs are most common in rocky landscapes where there is flow along buried rock surfaces. The interception of water movements by humans will be discussed fully later, sections 6.3 and 6.4.

Shine, the sun's energy, provides the energy input needed to transform liquid water into water vapour, then clouds and eventually rain, which in a sense is a fresh start with fresh, clean water. This, though, does not occur always, as some rain has some salt from the sea. As well, as it is falling through the atmosphere rain collects other things, for example, industrial pollutants, which give acid rain, or nitrogen-oxygen compounds formed by the energy of thunder storms. These compounds can actually be a significant contributor of nitrogen for plant growth.

The hydrologic cycle and salt

Most water is in the oceans, from where some of it is evaporated by shine, to cycle through and over the earth. Sea water is salty, around 3% sodium chloride, but rainfall is nearly pure water, except near coastlines where it is contaminated by salt, that has been added in aerosols blown from waves. Even when river water begins as relatively pure, the runoff over the land continuously collects significant quantities of salts, that originated in rocks and minerals, and which end up in the sea. So, the salt in the sea comes from the land and the salt on the land comes from the sea!

Though the water is generally very pure at the point of cloud formation, when the clouds form near the sea surface, sea spray is picked up and when these clouds move on to the land the rain falling on the land may contains salts. Depending on the rainfall characteristics these may accumulate on the land, and over a long period of time affect plant growth. For instance, the Murray-Darling River Basin of southern Australia has been calculated to receive in rain an additional half to one million tonnes of salt annually. Even though this isn't much each year on any piece of land, over the millions of years past it has added up, and how this is managed must be telling in the long-term use of the land in the river basins by humans.

Especially in areas where rainfall has been insufficient either in quantity or pattern of fall to leach it into rivers, much salt may remain in the landscape and become a problem especially when the hydrological balance is disturbed by human activity. This classic problem is common in southern Australia and many other semi-arid areas of the world, where it has developed following agricultural development, that has reduced the total water use, evapotranspiration. If deep-rooted native vegetation is replaced with shallow-rooted crop and pasture species that grow only during part of the year salt may be redistributed. Even though rainfall is small, drainage can be increased, water tables can rise and bring salt to the surface, especially in lower parts of the landscape. Production is reduced and in extreme cases is lost completely, resulting in bare ground that is subject to erosion and the exposed salt is readily mobilised to contaminate stream flow.

While some plants can tolerate high salt concentrations, no crop or pasture plants have well-developed salt tolerance. An important reason is that plants survive salty environments by either actively excluding salt from the roots, exuding it from leaves or isolating it in parts of the plant. Any of these processes consume photosynthetic energy and, therefore, reduce growth potential.

Some people dream of using clever new gene manipulation techniques to breed plants for salt tolerance. However, many people question the possibility of widespread application as it is not a solution to secondary salinity and such plants will most likely not only have low yields but also make the salt problem worse as time passes. The best solution lies in management that can change the hydrological balance to return the salt to depth below the root zone, or remove it from the landscape. In some cases, an appropriate solution may be to find a way to hasten the movement of the salt to the sea, or gather it into manageable lakes, and even breeding sea fish in them, as along the Nile River in Egypt.

2.9 Nutrients in soil water

Living things have evolved on earth by successfully 'fitting in' to these water systems. A plant is anchored in soil, accessing water that is held between the pores of soil. In a sense this is not much different from hydroponics, water is still the medium for plant nutrient carriage and supply. The water in the soil will have dissolved some of the accumulation of chemicals in the soil that happen to be useful to the plant, like phosphorus (P), nitrogen (N) and potassium (K).

2.10 Soil fertility

The dictionary definition of the word fertile is fruitful, being able to produce abundantly. We use the term fertilisation for both plants and soils, which may cause some confusion, because the actual process of making fertile is very different in these two cases. For plants, and animals, it is the process of the female part receiving pollen or sperm and so producing seed or offspring. For soils it refers to adding to the soil in an available form more of the things a plant needs so the plant will grow more.

The idea that soil was fertile, that is, productive, is no doubt very old in human history, long before the chemical basis of plant growth was understood. In our present use, the term soil fertility combines biological, chemical and physical properties. Organic matter plays an important role as it improves both water and nutrient-holding capacities and good farm and garden managers seek to maintain or improve both of those soil properties. It is important to remember that organic material does not supply nutrients direct to plants, as these must be in an inorganic form and pass into solution in soil water. So, we must never forget that what high fertility really means is that the soil has sufficient soil water which has dissolved in it sufficient of the essential chemical elements in an inorganic form to supply the plant needs.

Plants require about 15 nutrient elements and all but two (C and O) are obtained from soil. All the others, except nitrogen, originate from the rocks or sediments from which the soils are formed. N, the nutrient required in the largest amount, is very different as all N in natural soil-plant systems originates from the atmosphere, which is 80% N. Most N-gas that has been transformed into N-nutrients has been fixed by a variety of micro-organisms, which often grow in association with plants. Some also comes in rain as the spark of lightning may cause some oxygen and nitrogen to combine, or from human-managed industrial systems which simulate this, thus making nitrogenous fertiliser. Of the natural fixation systems, the combination of the plant group known as legumes with rhizobium bacteria is the most important. The legumes divert some of the products of their photosynthesis to the bacteria that live in their roots and accept nitrogen that the bacteria fix in a sort of exchange. So, even this process of gaining N from the atmosphere is, ultimately, supported by shine. That most N in the world exists in gaseous form in the atmosphere reveals that it is most stable in that form and that a lot of energy is required to extract it and combine it with other elements.

Individual nutrients are required by plants in different but balanced amounts. A useful classification is into macro- and micro-nutrients, depending upon the usual amounts found in plants. All are important and play specific roles in metabolism. A deficiency in any one will slow rate of growth. No amount of surplus of others will make up for it. On old analogy, less relevant in the days of hoses and pipes, is that of a water pail, like Jack and Jill went to fetch. This would have been made of vertical pieces of wood glued to a base, and the capacity of the pail would be determined by the shortest piece of wood. Similarly, plant production is limited by the nutrient in shortest supply, regardless of the amount of others present.

Of the macro-elements much attention focuses on P and N, which are fundamental elements in growth, P especially because of its compounds being involved in energy transactions, and N because it is a component of protein. Both 'cycle' through living things, and versions of the phosphorus and nitrogen cycles are given as Images 14 and 15. These images show the importance, and the complexity of the interactions of these nutrients.

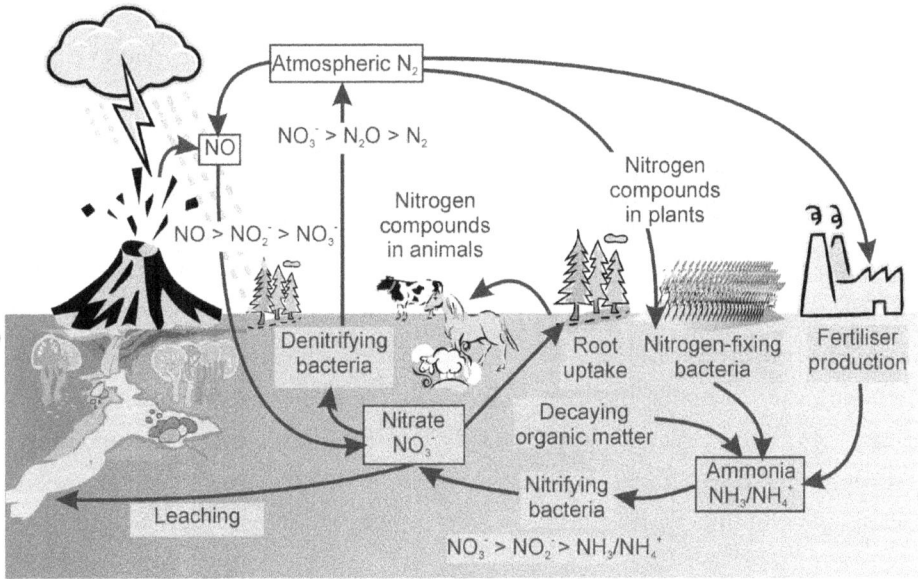

Image 14: The nitrogen cycle of agricultural production. This shows gains, losses and soil processes.

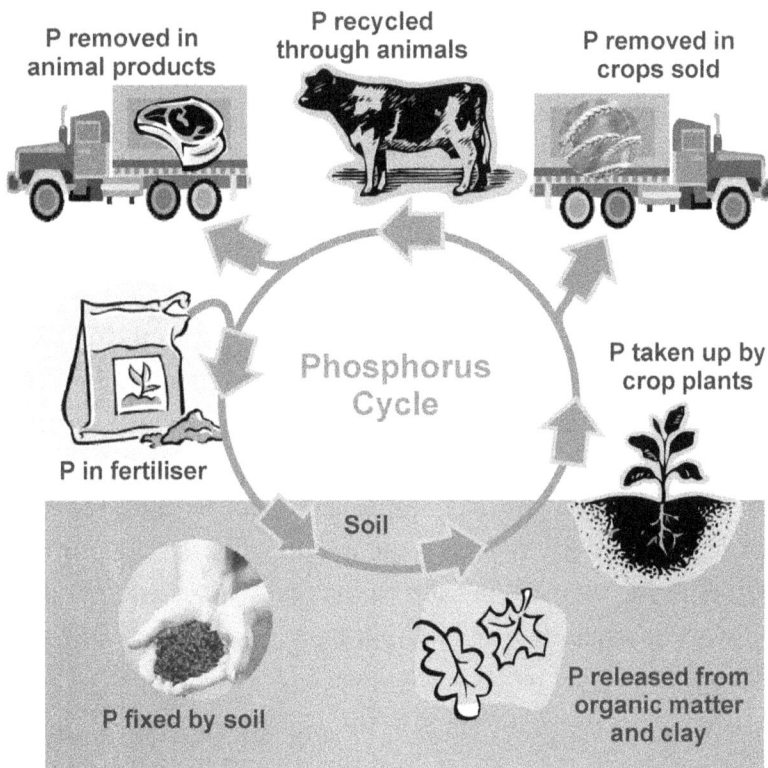

Image 15: The phosphorus cycle in farming systems.

Without human intervention most natural systems are closed, so that nutrients are recycled through the soil by litter fall, root decomposition, foraging animals and decomposing organisms. Though, water flow and animal roaming may move some from one place to another. In this case, the level of biomass production and biological activity has reached an equilibrium, albeit at a low level. Crop production also can be largely recycling, especially grazed as forage, but as higher yields have been demanded to provide for growing populations, harvesting removes more and more nutrients from the site. On some very favoured soils the weathering of parent material may almost replenish the supply of nutrients, but generally fertilizers and rotations must be used to maintain production at satisfactory levels. Understanding this is important in the consideration of production management systems, Chapter 8, including those described as organic, section 8.6.

Our relatively short life span makes it easy to forget that the processes of rain falling on the soil surface, sometimes running off, sometimes soaking in and seeping through the soil, eroding, depositing, have been going on for millions of years, even tens of millions on some sites. When cycling strongly, a lot of material may be carried to the sea, both in suspension and solution. Some very soluble compounds like common salt may be present at all parts of the cycle, and accumulating especially in the sea. Some, such as phosphorus compounds may be washed down and enter the food chain in the sea, from small organisms, to fish, to birds. If these birds gather on off-shore islands, sensibly away from predators, there may be massive deposits of guano, which is high in phosphorus. The cycle may be dead-ended, with the major land mass edging lower and lower in phosphorus. It is suggested that this has happened in Australia, with P accumulating on off-shore islands, such as Nauru and Ocean Islands, located in the Pacific Ocean. With P so essential for life processes, the low P on the land mass can have a profound effect on the level of biological activity, including plant life. So, P has been returned to the larger land masses as fertiliser, completing the cycle and allowing for great increases in plant growth. It is increasingly important that we intercept flows, for instance, massive amounts of P in sewage water, and return it to the cycle.

We humans have called soils as fertile, if containing a large accumulated store of plant nutrients, and set about exploiting them. But, for most soils, in the long run, this can be a snare and a delusion. In essence, we have simply seen this as an opportunity to mine the accumulated store of nutrients. In more recent times, science has enabled more exact measurement of nutrient levels in soils, and defined precise ways of maintaining these levels. The ways in which this is done will be covered in section 8.9.

Take away messages

1. Plants require a continuous flow of water for growth and survival from one fall of rain to the next.

2. Plants use soil water with nutrient substances dissolved in it.

3. Transpiration is the process of movement and transfer of water from plants, from pores in the leaves, to the atmosphere.

4. Evaporation is the transfer of water from soil to the atmosphere.

5. The relationship of the soil particles to water, and the capacity of the soil to store water is of enormous importance in the growth of plants and management of crops.

6. The Growing Season is the time period through the year when there has been moisture in the soil available to the plants, and temperature is not adverse for growth.

7. It is useful to look at both short-term and long-term trends in rainfall, to assess what plants we can expect to grow at a location, and as a base to irrigation plans.

3

Climate and weather

Perhaps the simplest way to define rain and shine is to say that climate is long-term weather, and that weather is the day-to-day expression of variations in atmospheric conditions that occur at any site. Data for atmospheric conditions include rainfall, sunshine time and wind speed. Also, is the climate fixed for a given location? And if it is changing, is this a natural cycle or is change accentuated by human activity such as carbon dioxide emissions? To answer these questions we need very long-term records, and not only for humans so recently arrived. That is, we need climate data on a geological and ecological time scale. However, the notion and technology of precise measurement and recording is a very modern one so we simply do not have enough direct data to be absolutely certain. Whatever, it is prudent to avoid those things that our scientific knowledge tells us have the capability of interfering with natural trends and cycles.

There really is a vast array of rain and shine combinations and we will touch on the major ones, but focus on the one prevailing in southern Australia, a relatively dry type climate. Compounding them are an infinite number and variety of niches, influenced so much by where the sun is at the time, and aspect (slope) and shelter, both from wind and direct sun. A good starting point is to be clear on 'where the sun is', that locally appears to be at the time. In places like the tropics near the equator, the sun is overhead twice a year and the seasons are less sharply differentiated. This pattern of relationship to the sun broadly causes air circulations and then rainfall patterns, so climate is a product of the shine pattern and rain occurrence. Then this is overlaid on the earth's shapes; seas and coastlines, mountains and plains.

3.1 Climate change

Climates are considered to be hot or cold, and wet or dry, either through the year or during particular seasons. The distinction is important to agriculture because climate determines the range of appropriate farming systems while weather challenges the farmer to maintain the productivity and sustainability of his chosen systems in the face of variations. Plans are made on averages, but it is important to remember that the average is derived from many observations, and the data can include some great extremes.

There is, unfortunately, a great readiness by amateurs to attribute all extreme events to climate change. July 2006 was a very hot and distressing time in England. In fact, it was the hottest since 1906. August was cool. However, the summer was, overall, one of hottest one per cent since recording began. The severe drought and reduced stream flow in southern Australia in the early years of the 21st century is publicly attributed by many to climate change driven by global warming. However, running means of rainfall show the 1890s and 1940s were drier, and scientists suggest changes in land use, especially with emphasis on reducing run-off, have had a huge effect on stream flow. Even without lower rainfall, the traditional use of large storage dams to supply urban water to coastal cities could be in serious difficulty. While we will only know for certain as time passes, it is prudent to make changes now. All of these issues have great community significance.

3.2 Climatic zones

Climatologists have mapped the lands of the planet in climatic zones, and any print or on-line atlas shows and explains these. Each has periods of unique combinations of rain and shine and it is worthwhile to discuss one zone as an illustration of this. The Mediterranean-type climate well illustrates shine variations and rain's contribution to soil moisture management. It has characteristics like those prevailing in countries around the Mediterranean Sea: a clearly defined and fairly reliable rainfall season in cool months, winter, with a generally very low rain occurrence in the warm or hot season, summer. In the Mediterranean itself, much of the region is sea, with a ring of land, including Greece, Italy, Spain, Portugal, Morocco, Algeria, and Libya. In California mountains restrict the area showing these characteristics. In the southern hemisphere, and especially in south western Australia, there is land at that latitude, and, even more pertinently, the land is fairly flat. The winter air circulation is from the south, over sea, and therefore cool and moist, whereas in southern Europe these air masses flow over land, so are cold and dry. Furthermore, in Australia any ranges are of modest height so the moist air masses from the Southern and Indian Oceans moving inland are not much elevated, carrying moisture a surprising distance inland.

A number of other regions, scattered across the planet, have Mediterranean-type climates. These represent a very small but important part of the agricultural areas of planet earth. Variation within them is considerable, as shown from the data for rainfall in Table 1, and to a lesser degree for temperature, Table 2.

These relatively broad zones also show variation at local level, for instance due to differences in topography, as a south-facing slope will get more rain, but much less shine than a north-facing slope. The warmer drier northern slopes often have distinctively different vegetation from the cooler, moister south-facing ones. Under livestock grazing this may be further compounded, as, given access to both in the field, animals will prefer the warmer slope in winter, especially for resting, so depositing (concentrating) nutrients there and making the site ideal for the growth of some annual plants which thrive on high fertility, such as cape-weed.

Table 2: Average monthly maximum temperature (°C) in Mediterranean-type climate regions. These are classified in two main types, hot summers and warm summers.

Seasons		Summer				Autumn			Winter			Spring		
Months	Northern (N)	J	J	A	S	O	N	D	J	F	M	A	M
	Southern (S)	D	J	F	M	A	M	J	J	A	S	O	N
Hot Summers													
Valencia, Spain	N	27	30	30	28	24	20	17	17	17	19	21	23
Los Angeles, USA	N	26	28	29	28	26	23	20	20	20	21	23	24
Perth, Australia	S	29	31	31	30	26	22	19	18	19	20	23	26
Warm Summers													
Porto, Portugal	N	23	25	25	24	20	17	15	14	15	17	18	20
San Francisco, USA	N	20	20	21	22	21	18	15	15	16	17	18	19
Cape Town, South Africa	S	25	26	27	25	23	20	18	18	18	19	21	24

Although climate can be defined by long-term rain and shine patterns, within the short term the weather may show huge variations. The Melbourne city, located in south eastern Australia, can be in the path of polar outbursts and also hot air masses from inland Australia. So, it is claimed to be the only place on earth where one can experience all four seasons in one day! Temperature falls from around 40 °C to 15 °C in one day have been known to occur. In farming this can be catastrophic, as, while freshly shorn sheep are quite tolerant of low temperatures, their metabolism cannot adjust quickly enough to the rapid fall in temperature, and many may die.

On the other hand, though at times inconvenient, variable weather is much more interesting than day after day of the same, and as a never-failing conversation starter. Perhaps, even more importantly, it allows dynamic life plants, and other creatures, are opportunistic, to seize niches created by variability. This is especially true of things like annual plants and insects that have huge numbers of offspring and mechanisms like seeds and eggs to avoid adverse conditions. The different behaviour of annual and perennial plants will be discussed later, in sections 4.2 and 6.14.

3.3 Combinations of rain and shine

Much shine and much rain

This is the tropical scenario of heavier showers of shorter duration, with plenty of shine, especially early in the day. This means the weather is hot and humid, with low evaporation and transpiration, often uncomfortable for humans, but great conditions for organisms such as fungi and some insects that cannot cope with the low humidity or some other combinations. We use crops and fruits that can flourish under these conditions and these do not have a specific 'dry' requirement for harvest.

Such tropical climates are good for crops like pineapples and bananas, Image 16. but can also support a range of annual crops such as haricot beans, and dense vegetation of crops and weeds, Image 17.

Image 16: Bananas, Congo.

Image 17: Crops and Weeds, Congo.

Much shine and very little rain.

The deserts of the world are so defined: long, very hot, sunny periods, rain spasmodically; at the extreme, even years without it. All of the continents except Europe have such areas. Remarkably, life forms have evolved to adapt to these conditions, including plants which take only a few weeks to germinate, flower and set viable seed. So, one big rain event may leave enough moisture for a complete life cycle. Seeds usually have dormancy, so failures do not lead to elimination of the species. Some insects and reptiles have also established amazing survival mechanisms.

Also, a number of range-land perennials are able to survive in this type of climate. For example, Image 18 shows the shrub vegetation near Broken Hill, New South Wales. Humans have been attracted to making these deserts bloom by irrigation, and these efforts will be discussed in sections 6.3 and 6.4, Part II.

Image 18: Desert vegetation near Broken Hill, New South Wales, Australia.

Much shine and modest but reliable rain

This is the normal situation in the Mediterranean type regions discussed above, really, defining them, by much shine and little rain in summer, plenty of reliable rain in winter. At higher altitudes it is snow, not rain. There are definite seasons, and living in such regions we are usually set up to cope, even to make a positive virtue of such periods. It's beach weather! Let's go skiing! The spring flowers are wonderful!

We grow crops that ripen, that is, to translocate their nutrients to seeds, as the moisture in the soil is becoming exhausted in early summer and the first bursts of high shine, with little rain, which is good harvesting weather for cereal crops. We grow fruits, often using water supplements, that ripen in summer to a high sugar content, for flavour in table fruit, or for fermentation, as in wine grapes. There is often enough winter rain to cause run-off and we intercept the hydrologic cycle to store water in dams, using it to ensure proper ripening of these fruits, and for the day-to-day household needs. We also like to look out on greenery such as lawns, accentuating summer contrasts, so we use stored water for that, setting up systems to regularly apply water to them. We like to play on nice turf during the fine weather, so we water lawns for games.

One of the paradoxes in life is that we have an interest in a variety of plant outcomes: we are not always interested in yield as expressed in seed heads. We only make the lawn grow so we can cut it and have nice fresh growth the right shade of green. We have an interest in keeping the plant from responding to stimuli, like day length or temperature, and running to seed, rather, remaining fresh and green and attractive and digestible for grazing animals. The flowering and seed setting stage may mark the beginning of the end for a plant's usefulness as a grazing plant: in the case of a plant like barley grass, which has a sharp awned seed head, it is the end. Where we aim to harvest the seed for our own food, or for our animals, as in cereal grains, we have less interest in leaf and stem growth other than as a prelude to transferring nutrients from these organs into a seed head. Hence the study of plant growth at different stages, called phenology, and their interactions with rain and shine is fundamental to maximising these outcomes.

In the Mediterranean type climatic regions we grow crops while the soil is moist. That is, plants are able to take up water from it. Through some winters there might be so much rain as to hinder plant growth, but we tolerate so much rain as it wets the soil and causes run-off, which we often capture. Also, catchments are at times declared off-limits for other activities.

We average out variations of these periods, based on past knowledge, and act accordingly. In the southern hemisphere the less shine and more rain regime usually begins in the March-April-May period, which we call autumn. Then conditions are very good for growth through the September-October-November period, which we call spring, when there is moisture from the winter, as well as continuing rains, and the shine is increasing.

3.4 Drought

We build a full pattern of life around climate expectations. It is when the weather deviates from the expected normal that we complain, as we can get into serious trouble, for instance, if it is too long a warm, dry period after too short a cool, rainy period. We call this a drought, normally defined by impact on crops and pastures, and to a lesser, though probably increasingly, by impact on water storage for irrigation and city use.

It is worth noting that the converse also upsets farmers, like a longer wetter, cool period with water-logging of some crops, more rain running off the surface, and flooding some land, which may even cause erosion and deposition. It is worth noting that this is not really the opposite of a drought. We don't really have a definition of that, perhaps an unusual situation in dryer climates.

Drought is a comparative rather than an absolute term. For agriculture it can only be defined as a longer-than-usual period of lower-than-usual rain at that place, so that growth of plants is reduced or their survival threatened. In some places, and for some crops, that may be several days, but in others, and for other crops, perhaps several months. Vegetable crops suffer water shortage quickly but deep-rooted perennial plants may survive for a long time. Logically, it is not possible to have drought most years. That is, we must revise expectations, and alter the general pattern of land use.

The rainfall variation for Nagambie in central Victoria, Australia, is shown in image 11 and we have discussed some features, section 2.6. The driest decade on record was the 1940s. Yet soon after, in 1956, this site contributed to the flood of a life time, when the whole length of the Murray Darling Basin river system was in flood. One interesting feature is the higher rainfall in the late 20th century, which raises the question as to what is 'normal'.

Drought resistance

*Nature has evo*lved a range of strategies by which plants manage drought, including escape, avoidance, and tolerance, and these are not mutually exclusive. Short-lived, annual or ephemeral, plants can escape drought by completing their life cycle quickly and surviving drought as seed. The seed bank that develops under annual pastures in semi-arid areas is a good example of that solution. Some plants can achieve the same result by dying back to a small underground bulb or tuber with dormant buds.

Plants that live through drought may actually avoid all or most of the deleterious effects of water shortage by maintaining high internal water status. Plants can achieve this by reducing or suspending growth, by reducing transpiration through stomatal closure and some shedding of leaves. Plants that continue to grow during periods of water shortage usually achieve this by avoiding or slowing water loss, and may also have an ability to tolerate internal drying.

Obviously deep root systems, which enable access to more soil water, can be important. Plants and crops that survive drought live to grow again but while they are growing there is no magic solution to greatly change the systems and quantities of exchange with the atmosphere of water for carbon. The nature of plant growth dictates that it cannot proceed without this exchange to a very demanding atmosphere.

Much as humans have dreamed about crop plants that can grow with less water, in practice huge gains in productivity, that is, yield, have been achieved by improved field management, which is what we call agronomy. This includes making sure water penetrates the soil surface, is stored in the soil and that evaporation is minimised by soil cover. Irrigation practices will be described in sections 6.3 and 6.4.

There are some interesting paradoxes in discussion of drought. If we define a drought as a time of poor conditions for plant growth, the opposite must be a time of very good conditions for plant growth, that is, having rather moderate rains throughout the season that keeps soils near ideal moisture content during growth, with never waterlogged, no run-off, and then a cessation of rain for ripening. On the other hand, if we define a drought as a time providing poor run-off into water storages, the opposite is a time of much rain, soils full of water with excessive quick run-off into storages, which could do a lot of damage to crops and cause erosion.

The mass media deal very badly with this term and surround it with emotive anecdotes. Reporting rarely conveys a clear picture to their urban audience. Perhaps it really is too complicated. The seasons 2001-07 in southern Australia were an interesting challenge. The average winter crop harvest for the ten years before was 16 million tonnes, but 2001 was very good for crop growth, giving a winter crop yield of about 24 m tonnes, up 50% on the average. The year 2002 was one of very low rainfall, which saw a crop of only about 10 m tonnes, 60% of the 10-year average, but in the popular press comparisons were made with the year before, down to 40%, rather than the average. In the season 2003 rainfall was not high, but it was well distributed, keeping the soil moist but rarely waterlogged, so produced a crop yield of nearly 24 m tonnes. However, there was little run-off and many storages were not much replenished, so to urban people, concerned about water supplies, these were two drought years. Rainfall was again well-distributed for crops in 2004 and 2005, but run-off into streams was way below average and the reservoirs for urban water supplies were low. The year 2006 was then one of near record low rainfall, so after six years of low run-off cities and towns faced an acute shortage of water, even though there were three good crop years. Paradoxically, good farming methods that used moisture where it fell very efficiently further reduced run-off and made conditions worse for city water supplies.

A further complication is that with grazing animals the emphasis is on how much herbage, leaf and stem, is produced, and not how much seed, and the shortage of fodder is usually through winter with plenty of feed in spring. So, a season with moderate rainfall in winter, followed by a spring dry enough to severely depress crop yields, may be fine for graziers, but very poor for crops.

In practice to grow a really good wheat crop needs only a modest amount of rain to allow sowing, little rain in winter so there is no waterlogging, then copious rain in spring when the plants are strongly growing, followed by dry harvesting weather. A pasture needs good rain in autumn, enough in winter, but spring doesn't matter so much. To fill reservoirs, we need a lot of rain to wet the soil, followed soon by a lot more, so there is run off. When doesn't matter much, but it usually happens in spring.

All of this is complex enough, but there can be many variables in the rain timing and amount even in the same sort of climatic zone, and huge effects on the resultant possibilities for plant growth. Take the Mediterranean climatic zones of Californian and Australia. The former has sites with an average of 1000 mm of rain per year, but it falls over about three months, with a lot running off, so moisture for plants is not available for much more than that, and the other 9 months are very adverse for plant growth. In contrast, in southern Australia, large areas receive 400 to 600 mm average annual rainfall, which will usually fall over 6 to 8 months, with a low average rainfall per wet day, and with only 4 to 6 dry months. One important consequence of the latter is that a smaller percentage of the rain runs off or washes through the soil, leaching it. So, any chemicals in the rain, such as salt, may accumulate on the land, especially in the upper layers of soil. We can described this as an interrupted hydrologic cycle, also see section 6.5.

In summary, the basis of agro-climatology is to study the longest possible records, predict the most likely pattern and develop crop types and length of growth phases to fit both planting and harvesting requirements into expected patterns.

Take away messages

1. Climate determines the range of appropriate farming systems, while weather challenges the farmer to maintain the productivity and sustainability of his chosen systems.

2. The earth contains a range of climatic zones, and each has periods of unique combinations of rain and shine.

3. Three main types of climates in agricultural regions are; Much shine and much rain (tropical areas); Much shine and very little rain (dry areas); and Much shine and modest but reliable rain (Mediterranean type regions).

4. Drought is a comparative rather than an absolute term. For agriculture it can only be defined as a longer-than-usual period of lower-than-usual rain at that place, so that growth of plants is reduced or their survival threatened.

5. Plants can show a range of strategies to manage drought.

4

Plants

4.1 What is a plant?

There are a huge variety of forms, shapes and sizes, from minute unicellular plant bodies, such as algae, to plants with a strong woody stem, trees, which are often huge. Examples of trees are shown in Images 7 (river red gums), 19 (pine trees) and 20 (olives). The common feature of plants is the ability to capture solar energy. Most plants are, at least part of the time, anchored to earth, mostly what we call soil, but sometimes rocks, not yet broken down to soil, and sometimes in aquatic, floating in water. We have already explained the impact of their 'arrival' as an entity through capture of external energy and creating a supply of free oxygen which allowed the evolution of other organisms and creatures, section 1.4.

Structurally, these generally have some differentiation of parts. That is, roots, to anchor the plant in the soil, and to take up water with nutrients in solution, and an above ground section, usually differentiated into stems and leaves, that intercept solar energy and to make carbohydrates. They have special conductive tissue which enables the movement up from the soil of the solution of nutrients in water, which we call sap. Image 2 shows some commonly used plants and their structures.

Perhaps one of the most important things emerging from the unified knowledge about plants is that the growth of plants, and all life, involves an intricate array of organic chemicals based on carbon. So, this is an organic, carbon-based economy.

Image 19: Agro-forestry: pine plantation near Ballarat, Victoria, Australia.

Image 20: Old olive trees near Madrid, Spain.

4.2 Plant Evolution

Plants are a stage in the evolution of organisms on the planet with the conditions and attributes, we discussed above. These have evolved in a watery environment with the universal fluid, the base for their sap. These operate in the temperature range in which H_2O is a liquid, the lower part of the range. Though, these are damaged if the H_2O becomes a solid, ice. These utilise the fact that energy is needed when liquid H_2O vaporises as a way of keeping their leaves cool and in the optimum temperature range for biochemical reactions to occur.

Plants usually stand in soil, kept in place by differentiated cells forming roots, with these roots exploring the soil where there is water, and establishing a flow of it to keep the plant alive and functioning.

Plants have evolved into many different forms, and the plant kingdom can be variously divided. One division is into annuals and perennials. Annual plants begin from a new seed each season, while with perennials, even if seed is set, the old plant lives on for some years, in the case of trees, even for centuries. The surviving perennial plant may become quite dormant in adverse conditions, with buds that have many of the characteristics of seeds, bursting out into new growth, when conditions of rain and shine are right. When considering the merits of annual and perennial crops it needs to be remembered that an annual 'expects' to die, and live on through its seeds, so it translocates virtually all of the nutrients in its stems and leaves to the seeds. A perennial sets stores aside to live on, often in basal nodes, and sometimes in the roots.

Another division is by size and structure. Creepers grow along the ground or climb on other plants; herbs grow low near the ground but with a definite stem and leaves; shrubs grow with a stronger stem structure, but bushy, generally a metre or two high; and trees emerge with a solid trunk and a canopy of leaves.

A further classification is whether they are early evolutionary forms that do not flower, called gymnosperms, or more recent forms that do flower, called angiosperms.

Yet another differentiation is based on leaf form: roundish needles, like pine trees, or flatter leaves with veins. An interesting sub-group is the legumes, which at some stage in their history 'teamed up' for mutual benefit, as what we call symbiosis, with special bacteria that live in nodules on the plant roots and fix nitrogen from the atmosphere into nitrogenous compounds that they exchange with the plant for carbohydrate. This brilliant strategy, which is not yet fully understood despite much research, enables them to compete more readily in infertile soils. Using them in agriculture to access the abundant atmospheric nitrogen, instead of using fossil fuel energy to manufacture fertiliser, is a very significant conservation management strategy.

Then there is the major division into monocotyledons, seedlings having one leaf, of which grasses are an example (see Image 2), and dicotyledons, where seedlings have a pair of leaves, of which clovers (see Image 2) and beans and peas are examples. One fascinating group is plants, such as capeweed and sunflowers, Image 21, like daisies, which grow as a rosette, which is a round base with radiating leaves, along with a fierce ability to compete and suppress nearby plants by shading, yet setting huge numbers of seeds.

Image 21: A sunflower crop.

Within this group with rosette form, some, when flowering run up a tall stem and have brightly coloured flowers to attract insect pollinators. One clever example is *Echium* sp., which evolved in the Azores and Iberian Peninsula, a species having brilliant blue flowers on a tall stem, which are attractive to bees, a source of honey, hence called Salvation Jane in South Australia. Though, this is called Paterson's Curse in Victoria, because of its invasive capability.

Others plants are able to put out runners along the top of the ground, rooting at a node to form a new plant, or even spread by underground runners, which we call rhizomes, that emerge, turn green, and effectively become new plants. One example is kikuyu grass Image 2d, that evolved in East Africa, which can come up metres away, even the other side of a garden path or fence. In the right place, for instance, as the summer rainfall areas of northern New South Wales, it can be a very useful pasture grass, Image 22.

Image 22: Kikuyu grass pasture in north coast region, New South Wales, Australia.

Some plants are parasitic or saprophytic, that is, they do not have an independent system, but attach themselves to another plant, sometimes the roots in the ground, sometimes on trunks and limbs, for example, mistletoe. They rely on the osmotic pressure being different enough to have a flow of sap, complete with nutrients, passing into their conductive tissue.

The idea of a defined growing season has already been discussed, section 2.6, as it defines the times when a plant can access enough water to grow.

Different plants have specific requirements for time to grow, and thus match with a particular growing season.

4.3 Seeds and germination

Everyone will have seen seeds of some sort, and will have seen a seed 'bursting into life' and producing a new plant. The seed, of course, contains more than just the germ of life, as it has some food store to start the new plant in its life. The total packages come in many shapes, for instance, pods, round fruits and long fruits, which vary in type and amount of covering, have a big size range. These also have various appendages to assist dispersal to new sites, for example, to catch on animal coats. Because we use the food store of the seed package for our own food, we can sometimes recognise a distinct fruit in which the seeds are contained, like an apple, for instance. When the apple falls to the ground and rots the moisture from it, and the food supply, are able to give the new plant a good start in life. It is a good exercise to collect and compare a range of seeds and fruits.

We humans have selected and bred varieties where the fruit does not readily fall, so we can come along and pick the fruit. Then we use the nutrients of the fruit for our own nutrition, and usually throw out the seeds: in the case of the apple, we call it the core. In other cases, it is not immediately obvious that the seed contains a food supply, as the embryo and the food store are in such a tight small package with a tough outer coat. The pasture plants we group as clovers are a good example, which have seeds that are small, roundish, and hard-skinned. These packages contain, precursors, that is, organic chemicals that control the development of the new plant and differentiation of cells into shoots, stems and leaves. In a general sense plants are not so different from birds and animals in this regard, which is not surprising when we remember that they have all evolved from the same ancestors and have a great deal of the genetic code in common.

Germination

This is the process of an apparently inert thing, a seed, 'coming to life', to become a plant. A seed is actually a package containing the genetic information required to build a new plant, with some biochemical precursors for vital processes, and a small food store to get the new plant started. Seeds, appearing to be inert, are nevertheless able to maintain vitality for a long time, if kept dry and in a low humidity environment, while they carry on real, but minimal, respiration. They also may have a variety of mechanisms to impede germination until there is a probability of success, when moisture becomes available. These mechanisms include, impervious seed coats which gradually breakdown, inbuilt biochemical dormancy due to inhibitors in the seed coat or the embryo, which these also gradually fade with time, along with an innate response to temperature and light. As in nature generally, success is judged by the continuation of the species and not the individual seed, as many may come to nothing, or be consumed by creatures. Plants usually produce many seeds and many are lost in the quest for survival.

Quite frequently, the seed will not be close to the parent plant, but in a new location, either from deliberate human harvesting and planting, or its dispersal mechanisms operating. Ideally, it will be covered in soil or plant mulch, as some seeds are inhibited from germination by light, and will not readily germinate unless covered, though a few are the reverse. To germinate, that is, to begin to grow, the seed must take up some water, from the soil: it may have to wait for some rain.

Most seeds are programmed to lie in wait for the right conditions, in the combination of rain and shine, which together will give a good chance of survival of the seedling. Evolution has sorted out those plants that germinate too easily on the first wetting, so most annuals have some sort of delayed action timer mechanism, called dormancy, and then a requirement for a rain and temperature combination to trigger germination, even a day and night temperature combination.

One interesting example of this is the mix of plants that commonly occur in pastures in the moderate rainfall parts of southern Australia with a growing season around 6-8 months. Seeds of various annuals are present, typically rye grass, barley grass, capeweed, storksbill, silver grass and sub clover. Each has its own response to rain and shine. An early rain in slightly warmer conditions, say late March to early April, will cause a strong germination of sub clover, which, with its own N source, will grow strongly and may dominate the pasture that season. However, a late break rain, say in late May to early June, will result in grass dominance, as the clover will simply remain dormant, most of the seed waiting for the next year.

With moisture available, the seed takes up some, swells and begins a new existence. Many processes are initiated. The most visible is the beginning of the root, which we call the radicle. That goes down into the soil, soon having some side branching root hairs to explore a large volume of soil, by penetrating between soil particles, and taking up water from the soil, which usually has some beneficial nutrients in solution. Then the shoot pushes up through the covering of soil, breaking through the surface, at that stage, whitish in colour. A remarkable thing then happens: as shine falls on the tip of the shoot it turns green, because the precursors in the seed are able to form chlorophyll, a green pigment. This is the vital substance that enables the capture of shine, the sun's energy, by absorbing carbon dioxide and forming carbohydrates.

4.4 Staying alive

The early stages of germination are hazardous. Moisture initiates increased respiration using more of the reserves to support the initial growth of the shoot and root. Growth reserves are small and may be exhausted in pushing through the soil, in the quest to expose the seedling leaves to the sun and atmosphere and allow the root system to explore the soil for water and nutrients. Until the leaves develop chlorophyll, the plant remains dependant on seed reserves and until the roots become active in the uptake of water and nutrients, the seedlings are vulnerable to desiccation and lack of nutrients, especially nitrogen that is needed in large quantities for the synthesis of chlorophyll. So, seeds buried deep in soil risk exhausting reserves in seeking light and those that germinate on the soil surface run a high risk of desiccation. Germination is just the first step in establishment of a new plant and is really only successful when the germinating seedling establishes itself as an autonomous unit able to compete with other establishing seedlings, including those of competing species.

The seed will not be the only living thing in the soil, as what is below the surface is far more diverse than what is above, what we can see. There are many small animals, many insects and micro-organisms, and when there is food for these organisms, and moisture, the soil becomes a seething mass of life forms, possibly trillions of individuals per kg of soil. Some of these will be 'friendly' to the new plant, assisting it to take up nutrients. Some will be competitive for nutrients. Some will process chemicals in the soil into forms more usable by the new plant. Among the assistants are the bacteria that quickly 'team up' with legumes to exchange nitrogen from the atmosphere for carbohydrates the new plant will make using the shine.

So, the new plant will enter a world with both hazardous and helpful elements. We must remember that plants have evolved to handle these. However, when we cultivate, or change the soil, even just from having grown a crop the year before, we can have a big impact on this seething mass of life, and studies of the total soil, chemical, physical and biological, are giving clues as to how best to help the new plant as it arises from the seed.

One of the popular recent changes in grain farming has been to use what is labelled conservation farming, in which minimum tillage and retention of all plant residues, such as stubbles, results in a population of microbes that are not always friendly to young plants. Perhaps, coping with this is simply a task for our plant breeders, perhaps as we have selected plants that have not faced that challenge. There are various other possibilities, like rapidly developing roots that go deeper and partly avoid the densest mat of organisms at the surface of the soil.

In this way we have the greatest marvel after life itself, the new, young plant connecting the earth to the energy of the sun, and progressing to the independent phase, where the roots taking up water that has chemicals dissolved in it. Conveniently, many of them are things that the plant needs for its basic processes to operate its system, especially P, K and N, to build plant tissue and to carry out energy processes. The water, called sap because it is a solution of these chemicals, rises up the plant tissue through special cells, to deliver nutrients. Also, at the surface of the leaf the water is available to cool the leaves through transpiration, as the energy needed for a change from liquid to gas is taken from the leaves. This is the equivalent to our cooling by perspiration, or the use of an evaporative cooler.

The youthful plant is normally in a vigorous growth stage, fairly rapidly building up leaf area and a strong root system. Which is a desirable character, advertised by seed merchants, is seedling vigour. The plant needs a range of chemical elements for continued growth, and the roots must take up sufficient of these from the soil, of what are often called macro elements, sulphur, phosphorus, potassium, magnesium and calcium, and micro elements, iron, copper, zinc, manganese, boron and chlorine. The other major elements taken up are carbon, hydrogen, oxygen and nitrogen, mostly from the air and as water. Quite soon the new plant will be self-sufficient, becoming independent of the seed, with a root system taking up mineral nutrients in solution in water from the soil, and manufacturing its own carbohydrates and proteins.

4.5 The older plant

Plants have evolved to be very clever, especially the annuals. A perennial plant, one living over a number of years, can have a tough time in a long hot, dry, summer. That is, unless it has been able to establish in a niche where there is extra water, like a creek bed, or a place of very deep soil, or an oasis or near a smaller spring, or unless it is aided by humans adding water to the soil. Some have evolved very deep root systems, so they can explore a great volume of soil. Some have mechanisms for shutting down activity so that little water is needed. Also, some share with annuals the prolific seed setting, so death of the plant is not such a catastrophe.

An annual plant aims to get the business of flowering, fertilising the ovaries, and transferring some nutrients to the seed all over before the soil moisture is exhausted. In response to such stimuli, as day length and possibly temperature, the plant gets the message that a stage of the season has been reached, which is correlated with the season of moisture being available coming to an end. The growing points then change to producing flowers, instead of leaves, and these flowers are fertilised in various ways, then seed is set.

The annual plant begins what we call translocation, the moving of nutrients from leaves and stems into the seeds ready to support a new generation of plants, itself becoming just a dry shell. This process we call senescence.

Many annuals set very large numbers of seeds, for instance, one plant of what we know as capeweed, a member of the daisy family, had 5,000 seeds. Then these annuals often have mechanisms for scattering their seed to find possible new niches that might happen to be better. Some 'hitch-hike' on the wool of sheep, skin of cattle or even the socks of humans. Some are so placed that animals or birds eat them but they can survive in the gut and are passed out at new sites, nicely sitting in a pad of manure. Brambles, called blackberries in Australia, are a classic example.

In fertilisation of flowers there is another miracle: in most cases two plants contributes genes to the seed. So, while the new seed is like the old plant, it is not identical, as it can react differently to many things around it. It might have resistance to invasion by a microbe or attack by an insect predator; it might have a capability for slightly faster root development; it might also have a slight difference in response to light, day length and temperature, for flower initiation. Then in the vigorous struggle for survival it might have advantages over its parents and compete more successfully. Some plant groups are self-fertilising, which has obvious implications for stability, after its kind breeding, but also limits adaptability.

4.6 Gene re-arrangement

Humans have been influencing the arrangement of the genes of the plants and animals they have been associated with and 'used' these from very early stages of being humans. It was, however, the monk Gregor Mendel in the mid-1800s who made the first big step in understanding inheritance. He grew sweet peas and observed differences in colour between new plants, then deduced a genetic mechanism of inheritance to explain the results. This soon became the basis of breeding techniques, bringing some rationality to the understanding of variation in traits. Just think about the wide range of breeds of dogs and horses that already existed before Mendel's time and the range of cultivars of crops such as wheat, rice and maize, was just as great, although less obvious. Mendel outlined principles of genetic recombination that are now widely extended to improve yield, quality, and disease resistance of crops and animals. Mendel identified the existence of genes but later work by Crick and Watson identified their composition as a double helix of DNA, leading the way to genetic engineering.

Virtually all the plants and animals we use in agriculture have been modified by cross breeding and selection but now new techniques of genetic manipulation are offering great advances with less effort. Like many new ideas in human society, this has understandably raised disquiet. Now it is possible to use genetic markers, that are linked to particular plant characters, to speed the otherwise tedious process of inter-crossing and selection. Controversially, it is possible to physically move genetic material from one plant to another, or even across species and orders, boundaries usually closed to sexual reproduction.

In the wild, there is a constant turmoil of genetic change, with selection through survival of the fittest moving plants in new directions. As we have such a short history on earth we do not recognise this turmoil. The arrival of a new 'fitter' organism at a new site will always mean ecological changes, and may even cause the demise of some species, which is very much the stuff of evolution. Partly as a result of this, extinction is the normal fate of a species, probably slower than average on an isolated site, greater than average on a recently invaded island site. Sometimes it seems cataclysmic, as in the early Cambrian era, only to be followed by the famous 'explosion' of new plants and animals.

4.7 Plant nutrients

The plant nutrients in the soil, some of which are in solution in the soil water, can be taken up by plants. These originate from weathering of rocks or additions in rainfall, dust and lightning. In many soils the soil water does not contain enough nutrients for plants to grow vigorously and in others this situation is likely to prevail after a period of cropping. This occurs unless extraction rates by harvest are small, or the parent rock is breaking down rapidly.

In farming, the level of nutrients is generally maintained by use of what are called fertilizers, naturally occurring or synthetic mineral substances of particular chemical composition, or organic residues from plant remains or animal manures. A plant does not take up nutrients direct from any organic substance, as these must first break down into a soluble inorganic form. So, plants do not distinguish between nutrients from inorganic or organic sources. The important issues are the nutrient composition, the quantity of fertilizer added and the availability of the nutrient elements. Availability refers to the rate of release of the nutrient elements from the more complex forms that exist in the soil and fertilizer. We will discuss all aspects of this in section 8.9, Part II.

4.8 Biodiversity

Understanding sexual reproduction and the genetic basis of inheritance helps us understand diversity and evolution. No individual is exactly the same as another. These maybe alike, but never exactly the same, unless derived from the division of the one embryo after fertilisation. The mixing and jumbling of genes in sexual reproduction means that offspring resemble, but are not the same as, a parent.

Also, each new individual is a different combination of genes that will be expressed according to a changeable environment. In natural communities, slight differences that endow competitive advantage are naturally selected and individuals gradually change, that is, to evolve. The difference is not much from one generation to the next, but over 10 generations, about 250 years for humans, the differences can be considerable. Organisms with shorter life and reproduction spans, such as plants or insects, can change and adapt much more quickly.

Natural selection has then led to the emergence of new groups, which are better adapted to the environment. We have learned to call groups with similar characteristics species and recognize this specialization in the wide range of organisms adapted to previous and present environments, and many of them are still extant in our ecosystems. We call this biodiversity, but we need to remember that the measurement of this can be influenced by definition of new species, and in some studies exclusion of introduced species. The same forces that cull uncompetitive individuals from species also cull unadapted species as environments change. One of the notable features of the most recent work on genetic constitution of organisms is the revelation of how many genes are common to many organisms. The vastly different looking breeds of dogs are 99.9% similar genetically and humans share 98% of genes with Orangutans.

In any mix of plants or animals there will be some that are recent arrivals, that is   recently evolved or arrived. These probably became increasingly dominant because of their more suitable genetic make-up, while some others are on the way out, as judged by natural selection. Our problem is that we cannot know which is which!

This also raises the question as to whether genes are actually lost when a 'species', that is, a group of plants that some human has decided are a distinct group, becomes extinct. There may be some fascinating information available as we complete the genome detailing.

4.9 Classifying plants

To better understand biodiversity and the diverse arrays of plants that have evolved over the ages it is useful to classify plants. Humans make order of their world by classifying and naming things, with one reason being the need to use a word or two instead of a long description. Imagine how cumbersome it would be if every time we wanted to talk about a table as 'a thing with a flat top and legs or some other form of support at which we often sit'! So, we have developed ways of naming plants using a system called to order by the philosopher Linneaus, similarly to the way we give people two names. We classify plants into orders, that is, major types, and families, genera and species. The people who classify plants are called systematic botanists. Perhaps, unfortunately, some of them are too systematic as they are what we call 'splitters', who create new species based on minor variations, or changing long established names, so we lose track of mentions in older literature. The opposite, called 'lumpers', do not cause so much confusion. In fact, sorting out material, and establishing that a collection really is different enough to justify calling it a new species is painstaking and time-consuming work. It usually involves wide-ranging collection and detailed study of material lodged in herbaria. Be sceptical of the integrity of anyone who claims instant finds of new species!

We need to remember, too, that these plants are really just a large range of combinations of much the same genes, with, remarkably, a small number of different genes, in a particular combination, which confers quite specific abilities in adaptation. We are slowly moving toward a system where classification will be mostly based on DNA, bringing certainty and stability to nomenclature.

Take away messages

1. Plants in general have roots, to anchor the plant in the soil, and to take up water with nutrients in solution, and an above ground section, usually differentiated into stems and leaves, that intercept solar energy and to make carbohydrates.

2. Plants have evolved into many different forms, and the plant kingdom can be variously divided. One important division is into annuals and perennials.

3. Many plants produce seeds to produce the next generation, which contain both the germ of life and some food store to start the new plant in its life.

4. Key crop types of seeds include grains and fruits.

5. Most seeds are programmed to lie in wait for the right conditions, in the combination of rain and shine, which together will give a good chance of survival of the seedling.

6. Along with water, plants need nutrients especially P, K and N, to build plant tissue and to carry out energy processes.

7. Virtually all the plants and animals we use in agriculture have been modified by cross breeding and selection but now new techniques of genetic manipulation are offering great advances with less effort.

8. Biodiversity of plants occurs due to mixing and jumbling of genes in sexual reproduction, which means that offspring resemble, but are not the same as, a parent.

9. Classification of plants is important to help identify these.

5

Ecosystems

An ecosystem is the collection of organisms 'living together', that is, interacting in a shared location. In the case of plants, this includes both the air above and the soil below the earth's surface. Typically, an ecosystem includes animals, plants, insects and micro-organisms.

5.1 Living together

Plants have evolved on earth to fit into the multitude of niches presented by different combinations of rain and shine, and, quite remarkably, these are often tolerant of others and share space and soil. Images 17 and 18 show plants living together in tropical conditions in the Congo and in semi-desert conditions near Broken Hill.

We marvel at impenetrable scrub, comprising many species of trees, shrubs, creepers, herbs, or the amazing range of volunteer weeds in the home garden. These result from interactions with available moisture in the soil, and with variations in temperature and light situations. Some plants are adapted to bright sunlight, some are shade tolerant. The grouping is sometimes called a plant community, and the totality of living things at a site is called an ecosystem. Even though the plants appear to be tolerating each other, there can be fierce competition, for water, light, nutrients, and even the secretion of adverse substances. Plants may literally poison each other, and also poison insects and animals!

As well as the plants, all sorts of other things live together, as an ecosystem comprises the 'producing' plants, the 'consuming' animals and insects, and the 'decomposers', micro-organisms. The whole collection has a multiplicity of dynamic internal interactions. Exterior influences will impact on the whole system, for instance, more or less rain, more or less shine, and the arrival of a new organism, either large or small. In one sense the ecosystem is a vague notion, as there is no specific area or boundary, so units can be large or small. Sometimes ecosystems are described as fragile, meaning that they will readily change in response to external forces, but again there is no objective standard. Equally it could be said that an ecosystem that is easily changed is highly adaptable.

As the plant is opportunistically taking up water, including any chemicals in solution, the nature and amount of soil water is critical and the importance of all aspects of the soil is immediately obvious. We have earlier noted the influence of photoperiod and temperature, section 1.6.

The strong competition between plants, especially for water and light, is very interesting, with some unique adaptive mechanisms. A species whose new seedlings rapidly form a rosette, like dandelions, capeweed, thistles, and Paterson's curse, quickly suppress growth of nearby plants through shading. Then the new plants soon have a secondary advantage as the roots, fed by the tops, grow quickly and take up soil water containing nutrients to the detriment of others. To these that hath shall be given! If a later stage of growth is an erect stem, like most of the plants in the above list, the flowers and seeds are well clear of other plants, and so can be easily fertilised by bees and other insects. Then the plant can soon have a strong hold, as well as being positioned to have seeds blown or carried to new sites. Also, add in some dormancy of seeds, so that even at the new site they do not germinate until conditions are cooler and more likely to stay moist. Collectively, we have a powerful competitor indeed. Also, no plant is intrinsically a weed, as it becomes one through a human decision that it is not wanted in a site.

Insects, too, have evolved to occupy rain-shine niches, differentiated into many species. It is said there be more insect species on earth than of all other living things combined, and insects will inherit the earth! These occupy so many niches, so many varied habitats, and very frequently by quite intimate associations with other living things, especially plants.

It is almost impossible to measure the impact of the European bee on Australian ecology, as it is more aggressive than native bees, fertilising a different range of plants, so giving them in turn an advantage. Most insects are prone to hot dry conditions, liking things moist and warm, though at some growth stages they are more protected from injury, as eggs and larvae, like seeds of plants, are much more protected from damage by adverse temperature and moisture conditions. Also, like the chicken and egg, it is hard to say which comes first, the egg, the larvae or the moth. Larvae are specially adapted to eat succulent plants, and the moths cunningly lay eggs in sites where, when hatched, the new 'grub' can easily find good food. The larvae, usually slow moving over small distances, are an amazing contrast to the moth, which may make intercontinental journeys, to find new life sites. Then, with huge numbers of eggs laid per adult, a minor insect arrival may soon be a dominant feature of the environment. The spreading of insects and some annual plants have a good deal in common.

We try to minimise the random introduction of new species into our contrived managed ecosystems, by what we call quarantine measures, and with the characteristics mentioned above, this is a continual challenge. Different life forms, such as eggs or larvae or moths, need quite different strategies. Micro-organisms are so easily carried by other biological entities, even simply stream flow, and annual plant seeds have various hitch-hiking strategies, like grass seeds in the wool of sheep or socks of humans.

5.2 Climax vegetation?

Overall, there is a continuous sorting out, a matching of genetic expression and habitat. It was once widely assumed, without question, and still is by a surprising number of people, that the group of plants growing at any site, called the natural plant ecosystem, is the best that could ever be at that site. That is, the most suited to that rain and shine combination, and called climax vegetation. In the 1920s Arthur Tansley, however, considered by many to be the first great plant ecologist, challenged this. Also, a little thought confirms his view that this could only be so if seeds of all plants likely to be suitable were present, and this is never so.

Sometimes it is not so, because the site is new, like new alluvial soil deposited by a recent flood, or the soil forming after a volcano erupts. For instance, as the basaltic areas of Iceland and the Azores Islands and Western Victoria. In the last case, the fast-spreading grasses colonised it well ahead of the slower colonising tree species. Sometimes it is because some plants from similar rain-shine combinations on other continents have not yet arrived. As Tansley pointed out, so much depends on the spreading mechanisms of the plants. Some seeds are 'sticky', as they cling to clothes, animal skins and are dropped off all over the place. Some blow in the wind. Some are in fruit that is tasty or otherwise attractive for animals to eat, and they emerge in faeces at a different site, ready to sprout and use the manure as fertiliser. There are many which have evolved elsewhere, which are equally, even sometimes more, suitable. One dramatic example is the proven suitability of many tree species to the grassy basalt plains of Victoria mentioned above, including sugar gums from distant parts of South Australia. Another is brambles, blackberries, introduced from Europe and flourishing in Australia, with seeds eaten by birds that drop the seeds at new sites.

In some situations, there is a low level in the soil of some elements vital for plant growth, restricting the range of plants that can grow there, and limiting the growth of those that can. Humans, as gardeners or farmers, of adding certain nutrients may dramatically widen the range of plants that can grow at that site, or the balance of those already present. Many people in suburbia have experienced the pleasure of making a new garden, starting with a raw construction site, building up the soil and steadily widening the range of plants that can be grown. Agriculturalists in Australia have done that with huge areas of land.

5.3 Natural protection of plants

All sorts of interactions occur in ecosystems, and humans have long harnessed some of these to assist a favoured organism. The term Integrated Pest Management (IPM) has been in recent vogue, but more primitive variants have undoubtedly been used by humans throughout history, though in more recent times more emphasis has been on reduction in chemical use. Plants have evolved a range of mechanisms to survive attack by animals, insects and micro-organisms, the fungi, bacteria, and viruses. Thorns and spines deter grazing animals, hard cuticles deter biting and sucking insects, and toxic substances prevent invasion of fungi. Just as plants can evolve to survive attacks, other organisms can evolve to overcome the defences. Nature is replete with specialized inter-species dependencies and the result is a continual battle for survival, that has important consequences for humans' reliance on plants.

First, new plants may be introduced to a region without their natural enemies, and quarantine introduced to preserve this. Having few enemies is advantageous for crop species, at least temporally, but it causes great difficulties in the case of weeds. Plants held in check in their natural environment, may flourish in new environments, as seen in the luxurious insect-free growth of eucalypts in many countries outside Australia. Also, there has been the spectacular outbreaks of prickly pear and Paterson's curse plants in Australia.

Second, an understanding of predator-prey relationships can be used not only to solve these problems in plant introduction. The control of prickly pear by the combination of beetle larvae and a fungus is a prime example. This understanding can also be used to help control disease outbreaks in cropping systems. As noted above, for centuries the challenge was to find a way to combine natural controls of diseases or insects with limited basic chemical treatments. Later IPM has been sued to reduce dependence on pesticides. This approach has seen success in terms of economic and environmental gains in many crops. In the case of insect pests, farmers have to develop the distinction between the good species that prey on the bad species, and also the concept of thresholds. In IPM, the predators are deliberately introduced and encouraged by managers. There are obvious possibilities in more rapid genetic change.

Insects that eat plant are another factor in the success or otherwise of plants in any niche. Emerging larvae can wreck a sensitive young plant, eating the leaves and so destroying its photosynthetic capacity, or at least severely reducing its competitive ability. Then, in the evolution of plants there has been natural selection of plants with ability to resist the attack of insects. That is, sometimes a physical condition such as type of leaf surface, more often a chemical that gives a bad taste, or even is toxic, to the larvae. However, the uniqueness in the taste of fruits, of even different varieties, is no doubt based on small amounts of chemicals present: we can imagine insects having favourite plants, too.

5.4 Influence of animals on plants

Many animals have evolved on earth, mostly by being parasitic on plants, and also influencing the success of a plant in a niche. Those that feed on plants only are called herbivores. Plants can often repel these animal attackers, like insects, sometimes physically, as with prickles and thorns on stems, or to protect the seed with spikes, called awns, or even bury the seed, as does subterranean clover.

Plants can sometimes deter animals with nasty chemicals. For instance, many of the family to which the eucalypts belong naturally contain fluoroacetate, commonly known as 1080, a good defence against certain animals. This natural chemical is used by farmers to kill feral rabbits and foxes by mixing with some favourite food which masks the taste.

Other plants enlist co-operation of animal in their life cycle, or ensure their seeds are spread to new sites, with hitch-hiking by clinging to woolly, furry or hairy animal coats. Alternatively, the seeds contain an edible attractive fruit. This can have great advantages if the seed resists digestion, as then it has a great start at the new site, being embedded in a pad of manure.

Animals, too, have evolved to be secondary beneficiaries of rain and shine, with special niches. Many are herbivores and some of these are fussy, loving to nibble short, sweet grass, for instance, rabbits. Others, like goats, seem to eat any herbage, including prickly and thorny material, like blackberries. Like sheep and cattle, goats are ruminants with four stomachs, able to regurgitate food that has been in the stomach for a while and chew their cud. Ruminants make the pieces smaller and add a further dose of saliva, which contains digestive fluids. It is then passed through the other stomachs, overall, getting nutritive value from the much rougher plant material.

The ruminant process seems to generate methane, a rather potent Greenhouse gas, so some people campaign against meat production. Countering this, if the animal did not digest the roughage, other organisms such as termites might, then also produce methane. Such is the wonderful complexity of life! Researchers are investigating how to minimise methane emissions from ruminant animals.

Some animals are carnivores, predators on other animals, so getting the benefit more indirectly. Then, there are those that are not particular, eating both herbage and meat, which we call them omnivores: we humans are also omnivorous.

In various combinations animals can have a profound effect on the life of plants. Herbivores by selection, may eat one plant and leave another, so giving the other a competitive advantage. Carnivores may remove animals which could eat out certain plants, so indirectly affecting both crop yield and/or the competitive outcomes in ecosystems.

In summary, plants grow in response to rain bringing moisture and the shine supplying energy. They have evolved into an incredible array of mixes of species, dynamic, responsive to change, constantly shedding some losers and evolving some winners. Insects and animals have major influences on the success of plants. Since, through the process of evolution, humans have progressively increased their management for the growth of plants.

Take away messages

1. An ecosystem is the collection of organisms 'living together', that is, interacting in a shared location. Typically, an ecosystem includes animals, plants, insects and micro-organisms.

2. Plants have evolved on earth to fit into the multitude of niches presented by different combinations of rain and shine, and are often tolerant of others and share space and soil.

3. Plants have evolved a range of mechanisms to survive attack by animals, insects and micro-organisms, the fungi, bacteria, and viruses.

4. Integrated Pest Management combines natural controls of diseases or insects with limited chemical treatments.

5. Plants also enlist co-operation of animal in their life cycle, or ensure their seeds are spread to new sites, with hitch-hiking by clinging to woolly, furry or hairy animal coats. Alternatively, the seeds contain an edible attractive fruit.

6. Animals can have a profound effect on the life of plants. Herbivores by selection, may eat one plant and leave another, so giving the other a competitive advantage. Carnivores may remove animals which could eat out certain plants, so indirectly affecting both crop yield and/or the competitive outcomes in ecosystems.

Part II

How we grow plants

6

How humans use plants

After a very long period of plant evolution and the development of other, mostly herbivorous creatures, came humans, who deliberately managed ecosystems in a more and more sophisticated manner. They had, and continue to have, a variety of objectives in growing plants. The underlying aim is to optimise the 'fit' of plants into patterns of rain and shine.

To some extent early humans also managed some of the herbivores and ate them, becoming what we call meat eaters. Though we will not include animal husbandry in this book, there are a few important aspects of this management to be noted. When humans were managing to eat the fruits of the plant, be it grain or other types of plant organs, they learned to maximise the production and quality of these organs. However, when they co-opted animals into a food chain, they learned to manage some plants to maximise their value to the animals: we call these forage or pasture plants. For herbivores it is the stem and especially the leaf that is important, at the earlier stages of plant growth, before flowering. Indeed, with many plants the change to flowering, and soon senescence, signals a sharp drop in their usefulness as forage.

Also, from the wondrous flow of evolution arrived a specialised group of animals, called ruminants, with the ability to digest rough plant material. That is, to break down the material and absorb nutrients into their system, as described above in section 5.4.

Importantly, using plants in modern agriculture generally involves ecosystem management. Most of us have had, or will have, some experience of this, whether it is from a school exercise of growing some seeds, the nurture of a house garden, or managing broad farm lands for fruitful production. In the history of agriculture managing farms was at first rather crude, but has become increasingly based on precise measurement, and this precision can be used to reduce adverse impacts.

To make and maintain sustainable operations in agriculture across Planet Earth to feed a growing world population, we must meet major challenges. It is, then, important to define precise questions based on a clear understanding, making measurements, and collecting evidence. Also, testing hypotheses is the only way to proceed, and we hope understanding what makes things grow, and how we grow things will assist in your understanding and participation in this task.

6.1 Harnessing the solar power of plants

From Part I we understand that plants have, in a sense, a unique ability to adapt to rain and shine, the earth and the sky. More precisely, they can take up water that has mostly fallen as rain, to enter the soil, and then move it up to an action site in the plant. There, using energy from the sun, the water is combined with carbon dioxide from the atmosphere to store the sun's energy as chemical energy in carbohydrates.

Further, the legume group of plants have come to an arrangement with some bacteria which enables them to access the 80% of the earth's atmosphere that is nitrogen. The nitrogen is combined this with some carbohydrate to synthesise amino acids, which, in chains, form proteins. So 'equipped', the plants evolved to occupy an incredible range of niches on this often inhospitable planet, then to be the basic support for most other life.

After hundreds of millions of years of the evolution of all sorts of organisms on Planet Earth, in responding to rain and shine, humans 'arrived'. We need not be concerned here as to exactly how, where and when, as we will be mostly considering recent activity, with emphasis on the last few centuries, and how current activities relate to our knowledge of rain and shine.

There are often debates about exactly when and where major changes occurred, as hunter-gatherers to farmers, and collectors of rain-fed fruits to beneficiaries of floods, such as the Egyptians along the Nile and finally deliberate irrigators with increasingly complex engineering structures. The likelihood is that these changes were gradual, evolutionary, responding slowly to changes in circumstances and tastes. Even in the hunter-gatherer societies it is difficult to believe people were not intelligent enough to be a bit manipulative, and perceiving advantages, to encouraging more useful plants in more convenient locations. That is, better for themselves or for the animals they hunted, and to avoid killing breeding females.

Aboriginal burning of grasslands in Australia is a good example, killing the seedlings of trees and shrubs, preventing development of forest and ensuring game could be seen, especially when attracted to the new growth. Likewise, it is difficult to determine, and perhaps futile and irrelevant to debate, which came first, technologies and inventions, or farming. It was probably iterative. Throughout, parts of agriculture have been just as likely to develop new inventions or adopt new technologies from developing industrial activities as any other human activity, and this continues. Yet paradoxically, farming in some developing countries has failed to adopt some significant inventions, for example, replacing the heavy cast iron hoe with a lighter steel model.

Agriculture, a term we use to broadly cover crops and grazing animals and fruit and vegetables, can be seen as an ecosystem with a maze of interactions, which are heavily dependent on the solar powered green plants. Image 23 shows a summary of these interactions.

In fairly modern times the image of the rough drawling cowboy riding into town, or the hayseed farmer driving into town in an old utility truck, are just interesting caricatures. In reality, the bulk of production in the developed world is from units managed by reasonably well-trained people who employ smart technologies, matching the sophistication of city life. It is inappropriate to describe them, as many do, as conventional, because they are especially adaptive, possibly more so that their city cousins. The word 'commercial' is a good, fairly widely embracing descriptor for those committed to maximising production in the world-wide effort of feeding earth's billions of people, more than half of them living in compressed cities, often on a different continent from the farmer.

Sunlight

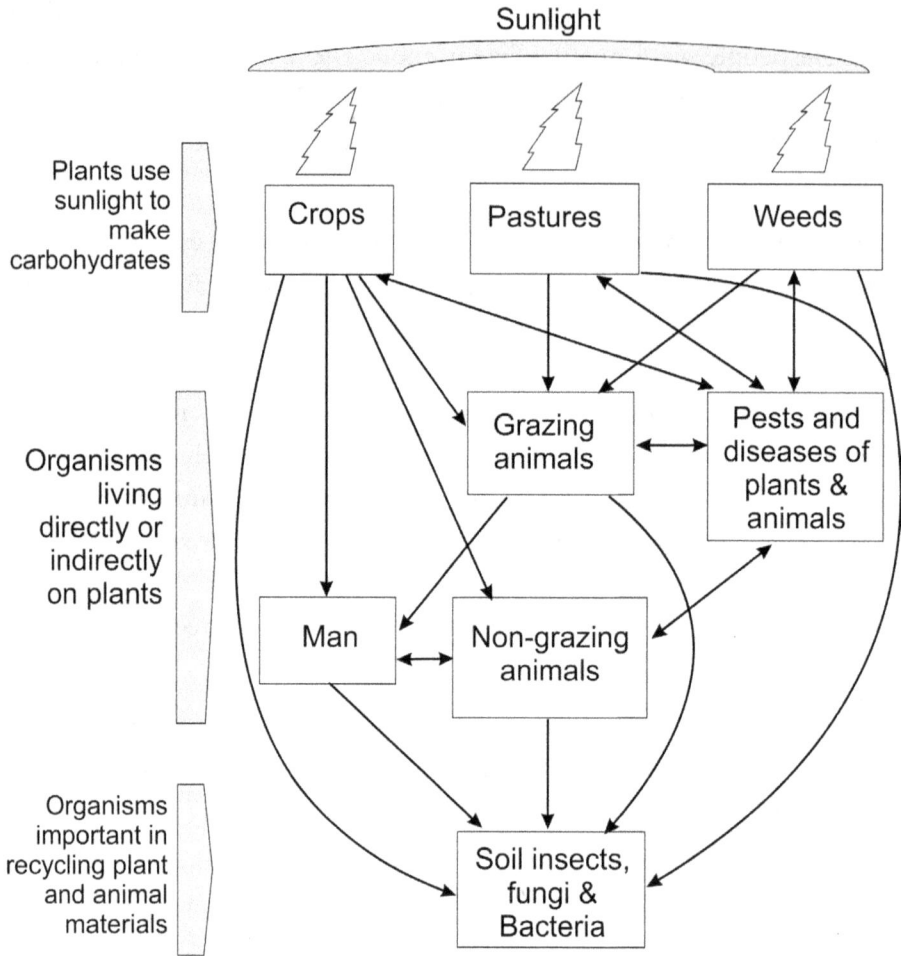

| Plants use sunlight to make carbohydrates | | | |

Crops Pastures Weeds

Organisms living directly or indirectly on plants

Grazing animals Pests and diseases of plants & animals

Man Non-grazing animals

Organisms important in recycling plant and animal materials

Soil insects, fungi & Bacteria

Image 23: Complex interactions in the agricultural ecosystem.

One of the useful ways of looking at human development is to consider the extensors of our human powers, which can easily be linked to growing plants and managing animals. We can identify three types of extensors: physical powers, of reach and power through leverage; sensory powers, of sight, sound, touch, taste and smell; and numeracy powers, of measuring, keeping records and calculating. In the first category we can imagine the gains from the invention of the spear, the bow and arrow, the lever, and the wheel, that have been incorporated into hunting of animals and tilling of the soil, with increasing efficiency. The second extensor, impacting much later, is illustrated in development of the microscope, which was a huge aid to improvement in the study of nature and the application of science. The third, partly covered by the popular term 'number crunching', covering record keeping, a huge variety of measurements and data analysis, was obviously very significant in improving management. Nevertheless, it was the linking of numeracy to the first two powers which ushered in the modern era with computer-linked operations. In modern agriculture there are many applications from monitoring the behaviour and growth of animals on distant ranges, to accurately sowing crops using machines guided by a Geographic Positioning System (GPS). Further applications are mapping yields while harvesting, and sorting fruit by colour and packing produce in the field.

In this context it is easy to understand how, especially with the harnessing of animals, small-scale gardening became larger scale farming. This then leapt in scale, with the use of engines powered by fossil-fuel. There are some who consider this the beginning of the end for Planet Earth, because larger scale farming and increasing human survival rates enabled humans to become a plague on earth having huge impacts on ecosystems. On the other hand, others put great value on the decreasing personal effort, in more efficient food production and less land area needed to feed one's family. This was a great step in improving the human lot, when education, culture and art ceased to be the preserve of the elite and became a rite of passage for all.

Though we may question the use of previously captured solar energy, usually called fossil fuels, we must still recognise the huge gains the internal combustion engine has made to return for human effort, and perhaps most dramatically, the large amount of land freed up when there were no longer draft animals to be fed, so more available for producing human food. We easily forget what it took to grow feed for the horse or bullock team, which was often nearly half the arable land area, and the limited work time imposed by tiring animals. We should remember this in the debate about biofuels, which effectively decreases availability of land for agricultural production. In using fossil fuels, it is perhaps the rate of use, and the processes used, that is the problem.

We can have an opinion as to the need for population control, but here we discuss how we can grow plants for supporting the population level, with growth rate as a given. However, what we must never forget is that humans, even in our highly developed society, are no less dependent on the land than were early humans, even if we are connected through a smaller and smaller specialised group of humans who manage the capture of rain and shine, and translate it through management of agricultural land into the needs of the majority. What has changed is that most people are more distant from this process and cannot, first hand, judge whether those entrusted with land management are acting responsibly and sustainably. Unfortunately, it is perilously easy for critics with little knowledge and scant understanding to be heard and believed.

Nevertheless, you can understand the fundamental components for supporting growth of crop plants, based on the evidence we have presented. We now associate the key processes and principles, described in the first part of this book, with the application of these in agriculture. This gives a basis for understanding planning changes to meet future needs.

6.2 Capturing shine

We can easily see how vital is capture of shine for all life on the planet and our dependence on it, directly or indirectly. However, we only capture a very small proportion through growing plants, and even this percentage is seriously reduced if the plant is short of water, or the soil is deficient in nutrients, or is eaten back by a predator, or sickened by a disease. So it is easy to understand why we choose a crop plant variety with the best possible genetic make-up, grow crop plants out in the open, away from trees, fence them to keep out animals, add needed nutrients to maximise growth, and spray the crop to control insects and diseases. Although we want to capture a lot of energy, the plant must not be 'all leaf' because it is often other parts that are useful: the trunk for wood, and the seeds or tuberous parts of the roots for food. We also maximise shine interception by orientation of plants, for example, in running rows north-south, and pruning. Then there is the shading of lower leaves so they become parasitic, as only about three layers will virtually capture all shine, which is important in row crops and forages. Our management must recognise all of these things.

The amount of food needed is determined by the number of people on earth, so, if we favour a system which captures a lower percentage of the energy and so has a lower yield, we will need a larger area of land to feed the people. In espousing systems, or resisting certain processes, we need to be aware of all of the consequences. This is an important consideration in proposing the total use of organic farming, see later sections 8.2 and 8.6.

6.3 Catching the rain

As a simple compound, water, H-O-H, arrives so simply, falling out of the sky in a fairly pure form. Yet, what emerges from our knowledge bank is that it is a very, very complicated task to manage water for the best in growing plants. It can run off the soil too much, and even cause erosion. But if we do not let much run off we are depriving our neighbours, or even our city cousins of their water entitlements. It can soak into the soil too much and displace all of the air, causing water-logging. It can also soak through the soil too much, leaching nutrients out of reach of plant roots, and even come back up elsewhere with salt in it. It can pick up and carry good, and nasty, chemicals in solution, and can carry other things in suspension. Disease causing organisms can move in it. It is a heat carrier, and in the plant it must pass up through and be available to cool the leaves. While moving up the plant to do this it must also carry essential plant nutrients.

When we grow plants we can intervene in all of the things above. So, water management is an almost mind-boggling task, and there is a massive amount of research into its use. We intervene to have it available at different places, through storages and canals and pipelines, and pumping up out of the ground. Also, at different times, storing run-off water, then using it later when rainfall is insufficient, for irrigation. Irrigation disconnects production from the unpredictability of rain, which allows the production of larger yields with greater certainty, and many people believe it is the only way to produce. However, there are physical limitations to where we can irrigate, costs may be high, and continually irrigating may lead to soil problems. Notwithstanding the spectacular side of dams and irrigation, and crops that simply rely on rain, which we call rain-fed crops, support a large proportion of the people on earth. The well managed rain-fed cropping, with inputs matched to the available soil water, serves humankind well. Rain-fed crops are shown in Images 3, 4, 21 and 24 .

Image 24: A canola crop, Wimmera, Victoria, Australia.

6.4 Variations in rainfall

Other than in some tropical regions, where in the rainy season it seems to rain every evening, more often than not the patterns of rainfalls vary. Even in the 'wet' season in southern Australia, we have fine days or wet days, light rain or heavy rain, and we quite like the variety until it inconveniences us! We establish agriculture in lands with variable rain, facing the challenge with various management strategies. However, in areas where rain is not completely reliable this leads to highly variable yields and, potentially, famine. Modern economies recognise this as 'risk farming' and use the storage of some of the harvest from the good years to use in the adverse ones, usually called droughts. This can be averaged over the years, to still allow very profitable agriculture.

Other aspects of rain variability have been discussed in sections 2.6, 3.2, 3.3 and 3.4. As we have noted, southern Australia is an unusual place in its rain patterns. Most notably rain is more spread through the year than other places in similar latitudes, such as the lands around the Mediterranean Sea, California and Chile. The soil is more gently wetted to some depth, giving a longer period when there is moisture in the soil and available to plants. However, this means that much more of the time the crop is on the brink of shortage, and that there is a lower percentage of run-off, especially when introduced pasture species have been planted to give more surface cover than existed naturally. Recall the hydrologic cycle, from sea to clouds, to rain, to land, to rivers, then back to sea, section 2.8. This is frequently interrupted, and salt that has come in the rain is usually not returned to the sea. This probably has been so for some millions of years, especially since the southern and central parts of Australia became drier. We believe there was once high rainfall and rain forest over much of it. Though, in geological time this drier time may only be for a brief interval, maybe a million years, as it is the time when humans have evolved and lived in the area.

In this management we need to understand how to capture rain into the soil by the ways we manage the surface, to hold what falls while it soaks in, but to prevent waterlogging. We need to minimise direct evaporation from the surface. There is a lot to know about water and soil and root zones, and much scope for improved management. This has been especially important in southern Australia, where there is a northerly long east-west arid boundary limiting crop production. Also, small improvements in any aspect of improved water use can mean a huge increase in crop area and/or total crop production. Plant breeding for drought resistance, for plants to use water more efficiently, has played an important part, especially when coupled with precise and high-performance equipment. This is used in perfect timing of crop sowing and fertiliser additions, to enable enviable levels of productivity from limited resources. For example, wheat yield per unit of seasonal rain has doubled over the last 40 years. By contrast, the yields of most crops over most of Africa have not increased in the last 100 years. A similar doubling there may well have averted most of the famines.

There is now much science in managing the capture of rain and optimising soil capacity for moisture storage, and modern farmers use a computer model to incorporate this into management practices, and also use crops that have root systems capable of using the full depth of moisture storage. For grain crops the ideal rain pattern is a moist autumn while there is plenty of shine for growth, a fairly dry winter when plants are not so actively growing, and plenty of rain in spring as shine increases and plants grow actively. This combination of rain and shine is especially important for crops to grow taller and denser, as then these use more and more water. Ideally follows a warmer, dry period for ripening and harvesting.

6.5 Impurities in rain

Of the many things that get picked up and carried by rain, salt is arguably the most important, though, if near industrial areas, acid rain may be a problem. At the right concentration salt is good for us, and trade in salt has been an important part of human history. However, too high a level can lay waste large areas of land, so rainfall or land-use patterns that cause this deserve some special comment here.

Accumulations of salt occur on, or in, land in most semi-arid regions. Witness the vast salt lakes of western Bolivia, with salt deposits metres thick.

Taking southern Australia as an example, the situation in the current time is of small rains per wet day over several months resulting in an interrupted hydrologic cycle. Though perhaps temporary in geological history, yet this has prevailed for long enough to leave its mark. The rain, carrying salt, soaks into the soil, wetting to a different depth each year, but over, say, 1,000 years there would be a most common depth, leaving a soil layer with most salt in it. As the plants take up water fairly much in balance with rainfall, the situation would be stable, but perhaps slowly trending towards movement down slope, especially in areas with wetter rainfall cycles and salt water in deep alluvial layers, with enough shine to evaporate the water and leave some salt lakes. So, over many millennia there may be an inexorable rise in salt levels in many catchments, with, eventually, more and more of the slopes being salt affected. The situation is sometimes aggravated by salt from ancient sea beds, as in the mid-Murray river near Swan Hill, in northern Victoria, Australia.

Changes in land use accompanied by plants with different root patterns, such as changing from trees to pastures and cropping, result in altered patterns of water uptake, with increasing water moving through the soil down-slope, as salt locations. It is important to understand that the total amount of salt may not change, while it changes in distribution. Ideally, the salt would be ultimately returned from where it came, to the sea, by having it carried in ground water moving down, or surface water flowing in streams.

Perhaps, rather than having a general strategy of reducing salt movement, it should be locally increased, by using pipes and/or narrow channels to segregate the salty water from the purer water, then in some places to gather it into local salt water lakes. These could be used for marine fish and/or seaweed production, boating, even swimming, or alternatively gathered and evaporated, to sell the salt and minor by-products, as is done along the lower Nile river in Egypt.

In other areas there could be actual completion of the cycle. That is, a direct pipeline could be used to take the salt water to the sea. This is because the river bed is best used for high quality water, and in any case may be long and winding with locks and other structures. Improved, cheaper pipes, combined with the prolific shine, converted to solar energy for pumping, may make this economically feasible. While solar energy could be used to lift the water over the ranges, some hydro-electric power might be generated in the run down to the sea. In any case, the costs of completion of the cycle might be the price we pay over the long term for having towns and cities and production of high-grade products for the people.

6.6 Knowledge-based water management

That we must fit our activities better into natural cycles is now well recognised. All water must be managed, not just the water we use to make plants grow. We must know the levels of all substances at all stages of the hydrologic cycle into which we tap, and we must know what we are adding, for example, in soaps and detergents as well as in industrial wastes. Then, there is a case for restricting certain components of products sold in supermarkets. Also, we must know exactly what will be in our waste water, and its impact. This is especially important for our vast cities, where people for so long have taken little interest in water beyond the toilet bowl and washing machine, and have used their nearby rivers and lakes and seas as dumping grounds for partially treated sewage, industrial washings and polluted storm water. We now accept that water must be left in a state when it can be used again. Readily available cheap pipes and pumps open up all sorts of possibilities.

Systems that appeared reasonable for a few people on earth have often proved dreadful when there were many more people. A few humans living fairly simple lives in a village had little downstream effect, as the environment was substantially able to deal with their burden on water movement. Perhaps the worst dangers were microbiological, diseases that spread downstream. As villages grew, excrement was kept out of the streams by various practices, such as the use of pans, carried away by the night cart man, whose truck used the back lanes, a peculiar feature of older suburbs. Then came what were rather quaintly called septic tanks, an apparently great improvement as the water closet (WC) could be inside the house. Also, wastes could be washed into a concrete tank located in the ground, where it underwent anaerobic breakdown. next the fluid flowed out into a long trench filled with gravel, for final aerobic cleaning, and soaked away, often into ground water and other unwanted places. Disposal in this way relied on the combination of a water supply, soil nearby capable of absorbing water, and no close rivers or streams, into which the water could seep.

At higher population density, and complexity of activities, came the sewerage system, effectively a dumping place for all of the unwanted materials from the city, both domestic and industrial. Operation was simple, to run it all down pipes to some point outside the town, filter out the solids, and allow some aeration of the fluid in ponds. Then pass fluids of varying composition into the sea or a river, with total disregard for the content of N and P in it. Sometimes the water was used for irrigating pastures, as in Melbourne's famous Werribee state farm.

As city life became more complex, for instance with washing machines and dishwashers, special detergents were used, often quite high in P and salts, and the management of the resulting water became quite challenging, especially as though the P and N increased the value of the water for plant growth, the salts reduced its value for plant growth. In the case of P, which is so critical for all life processes, and so is used widely as a fertiliser, the known deposits are becoming exhausted. So, it is imperative that conservation of P is practised. For N, again an essential plant nutrient, salvaging offered significant economic gain. The alternative process, to denitrify to N gas in the atmosphere is expensive and bad for climate change. In contrast, to use this N in productive plant growth can save substantial use of fossil fuel and energy in manufacturing fertilisers.

It also important to conservatively use plant nutrients to minimise runoff from crop fields. That is, high nutrient levels on runoff can cause environmental problems in nearby lakes, streams and rivers, in particular eutrophication.

Cities also include huge areas of hard surfaces, including roofs, roads and car parks, from which water, quaintly named as storm water, runs into drains, and then into rivers and the sea. It carries much of the city detritus such as paper, plastics, cigarette butts, dog excrement and plain ordinary dirt. So, city people certainly do not leave water as they find it, but rather in a much worse state in terms of dissolved, suspended and floating material. The challenge is to make cities water sustainable, so that, while these use water as a convenient carrier, these also accept responsibility for its restoration and the use of the nutrients in it. Using the effluent water to grow plants is an attractive part of this proposal.

Curiously, the system of water use on single country homesteads, or even small groups of them, has been quite different, virtually the obverse of the urban one. The farmer typically captured the water from the roof of the house and possibly sheds as well, and stored this in a tank, quaintly labelled rain water, that was laid on to the house for domestic use, namely for kitchen, bathroom, laundry, and possibly the toilet. Other water, whether from rain falling on roadways and yards, was collected into a dam, perhaps supplemented from bores, creeks and channels from major regional schemes, which were all considered lower grade water and used for making plants grow, washing down and often toilet flushing. Many new subdivisions have something akin to this dual system, and if cities had set up in this way from the outset we would now have a much more sustainable situation.

6.7 Having capable plants

The challenge is to have plants that 'make the best' of rain and shine. We can see that living things are very variable, including in their ability to utilise rain and shine. The clusters that we call species, and even varieties within a species, are very diverse. Just look at any group of people, such as a school class. However, though we talk a lot about biodiversity, when it comes to a crop in a field the ideal is as many plants as possible carrying the genes that will allow the plant to grow well. This means to fully as possible utilise that rain and shine combination, to resist attack by microbial diseases and predator insects, to use both the available soil nutrients and soil moisture efficiently, while producing a large yield of foodstuff for humans that is nutritionally good. At the same time, we must recognise that the exact genetic make-up that will deliver these outcomes is changing and elusive, so the scientists making the gene combinations need to be well informed and use the best technologies.

Looking back over history, we can imagine plant selection gradually being made more widely than just down the valley. Next, over the hills into the next valley, then farmers exchanging seeds with travellers, and finally importing seeds from other countries. One can speculate that people learned about the 'birds and the bees', noting that even an animal 'after the kind' of its parents was always different in subtle ways.

It was only relatively recently, especially contemplating the writings of Charles Darwin, that humans began to realise that plants varied in their genetic make-up in ways that gave some an advantage over others in natural selection, and that this could be harnessed to get better production. The monk Mendel growing and observing his sweet peas in the mid-1800s was an important event. Until then, it was mostly people keeping seed from outstanding individual plants in a crop or garden, or even in the wild, as having some special characteristic that was valued. These traits included flavour of fruit, or a benefit like resistance to attack by insects of disease. However, from Mendel's time, humans began not only to select, but also to intervene in the fertilisation process. Initially, by our standards, this was done crudely and haphazardly for re arranging genes, to get a better fit into niches for their plants and animals.

Specialists then began to appear, called plant breeders, who tried to ensure that the right genes were combined in the new plant, by shaking pollen from one plant onto the flowers of another. This was still rather hit and miss, being very slow and time consuming, as it involved doing this with hundreds of flowers. Though, it was very rewarding when it worked, on some plant varieties at least. Plants had so many genes, and, as Mendel had discovered, some genes were recessive, that is present but suppressed, while some others were dominant. While a plant only needed one copy of a dominant gene, from one parent, to express a trait, it needed two copies of a recessive gene, one from each parent, to express the associated trait.

Testing by a plant breeder as to whether a 'cross' had been successful meant growing the seed the next season and observing the plant, and in the case of disease or pest resistance, hoping the pest would turn up. Even then, the system of dominant and recessive genes made the assessment of genetic make-up uncertain and progress was slow. A type, really a sub-species, that is defined and used in organised production is called a cultivar.

At first, just flowering and having some fruit would have been seen as being a useful plant. Much more recently, prolificacy became valued, which meant maximum yield for human effort, or garden area, or more recently still, yield per litre of fossil fuel. So, came the concept of what we call yield. This was followed by the harvest index, the proportion of the plant biomass that ends up as the product humans want, like grain. So, instead of a leafy plant with little fruit, even if it was good at intercepting shine, came the emphasis on a plant leafy early, but then translocating as much as possible into fruit. Early crops would have had a harvest index below 0.3, later selections were higher than 0.5. This reflects a more efficient use of rain.

Against these gains must be set the reliability of yield. For millennia there was minimal storage so the annual crop was grown to feed the people through to the next crop. Then failure meant famine and starvation. Failure could be due to adverse weather, or to disease and insect attack. The selector/breeder needed to understand the risks and their probable occurrence. He or she also needed to work with people who understood management practices that might reduce the impact of adverse conditions. Gain in yield in any district, that is assessed to fit the trend line to long-term yield data, is a part of finding the right genotype and using the right farming practices that can cushion the risks.

6.8 Modern science-based farming

It is often suggested that the whole modern system, cultivars plus practices, is much more susceptible to adversity than peasant farming, section 8.4, in that the high yielding varieties require a lot of fertiliser. In a perverse way this is true. So, is an old variety that yielded say 1 tonne per ha in a good season without any fertiliser, and 0.4 tonne per ha in a bad year, better or worse than a newer variety that if well fertilised yields 7 tonnes per ha in a good year and 1.0 tonnes per ha in a bad year? Then, too, among modern practices we must include the farmer and merchant ability to properly store food supplies against a bad year.

Such breeding through flower fertilisation is more or less limited to that species or near relatives. In recent times newer tools have made it possible to choose the new genes with more certainty and insert them into the plant more quickly, and to try a much wider range of genes. Genes can even be transferred from apparently unrelated species such as bacteria. However, though species are not inter-fertile, they are related in that they have evolved from the same ancient common ancestors. So, it is a matter of if the genes fit, as the organisms are related.

The breeder can also assess success much more quickly and certainly by studying the gene sequences, and making selected changes based on this knowledge. When lay people understand this, albeit with strict controls, as there is in any new process, there can be widespread benefit. Surprisingly, plants produced by this process have become known as Genetically Modified (GM) plants. Some have demonised this, opposing it as something new, whereas intervention in gene combinations is as old as humankind.

6.9 Having the right range and type of nutrients

Adding the right nutrients to soil is to make the best fertile soil, whether by good luck or design. In practice, to a grower of crops it is many faceted, which at times is elusive, but when it 'works' very rewarding. A dramatic illustration of this reward is that, whereas farmers in Europe some 200 years ago sowed one seed and harvested as few as four, the modern farmer on the same land expects to sow one and harvest at least a hundred, and often two hundred seeds.

Clearly, there must be an adequate and available supply of all of the essential chemical nutrients for plant growth from the time of germination to the ripening of the seed. Then this must be maintained over years of cropping. Otherwise the level can slowly but inexorably fall in systems that do not completely replace the nutrients removed by the crop. Anything less than maintenance will lead to inefficient use of the rain, shine, and plant genetic make-up. Restoring nutrients need to be in a form readily available to plants, added in an energy efficient manner, and in a form not having adverse effects on the living soil. This is quite a challenge!

Another important aspect is the level, and nature, of organic matter. Organic matter was recognised as beneficial long before its composition was understood. An important step, in Europe, was analysing muck from in-door winter keep of animals, so that precisely what was being returned to the fields was identified. Then came precise soil analysis to exactly determine what was happening to the soil level, to maximise return from effort. This resulted in attaining the best yield for the rain and shine combination. The organic matter must have the right ratios of critical elements such as carbon, nitrogen, phosphorus and sulphur. The whole process needs many other elements in such small amounts that we call them trace elements.

We should also note that litter, that is, plant remains on the surface of the soil, is mostly pure carbohydrate in the form of cellulose, which is useful as mulch. Though, this does not entering into the 'life' of the soils, as does the soil organic matter, comprising C, N, S and P nutrients.

6.10 Using the living soil

Much of the organic matter will be bound up in living things. The soil is teeming with life, which we call the soil biota. These have active metabolic processes, which involve oxygen and nitrogen and, especially, carbon.

We can do a lot to affect the type and numbers of organisms. These can be involved in the availability of nutrients, the physical condition of the soil, including binding particles, and improving water flow. These organisms can even have impacts on some soil borne diseases, in sometimes reducing them, though sometimes increasing them. Waterlogging is bad for such life. Acidity is also an issue, which can be addressed by adding liming, though this has long been practised without precise definition of effect. Adding any substance or organism, including worms, for whatever purpose, like preventing disease, killing insects and weeds, is bound to have an effect on balance of activities. This applies to all so-called fertilisers, even bio-fertilisers, which are substances that include both nutrients and beneficial organisms that can start new colonies in the soil. Again, knowledge and precise measurements are necessary.

6.11 A carbon-based life

Recall that carbon is captured from the atmosphere, as carbon dioxide, in the process of photosynthesis. Understanding its place is profoundly important in overall management, so we must understand the carbon cycle. This element is ubiquitous, as found everywhere, including as carbon dioxide in the atmosphere, as carbohydrates as a main component in living things, and as hydrocarbons in coal, oil and gas. People who talk of the need for a carbon free economy are frightening in their ignorance, as in reality it cycles constantly through the whole of life and being. There is release back into the atmosphere as carbon dioxide, operating through both respiration of living things and chemical oxidation. There is increased capture of carbon dioxide through growing more plants and, therefore, through photosynthesis. Both represent energy flows. Soil that we consider fertile have a high level of carbon. Animals grazing plants convert sugars and closely related compounds into more complex carbohydrates, which when passed out as faeces will form an important part of the support system of the soil. When we plant a crop the preparation of soil can profoundly affect soil carbon, as tillage aerates the soil and increases some sorts of biological and chemical activities, mostly for the breakdown of organic matter and associated production of carbon dioxide.

Modern farming in countries like Australia, and to a lesser extent the USA, minimises tillage in preparation for a crop, even to the extent of directly drilling the seed into the soil without prior cultivation. This enables stubbles to be retained, to both minimise disturbance of the soil organisms and retain. moisture. Image 25 shows the stubble after a crop has been harvested in northern Victoria, Australia. The normal practice is to use narrow seeder points, often with the machine guided by a GPS, to sow the next crop in between the rows of the last one, with minimal soil disturbance.

These changes in practice have occurred mostly due to the application of science to improve the response of plants to rain and shine. Elements critical to the growth of plants, such as P and N, and C, have had special attention. C is conserved in organic matter as much as possible. N is captured whenever possible from the atmosphere by using a special group of plants, the legumes, which during their evolution have gained a truly astonishing advantage.

Raising the level of carbon kept in the soil can only come from growing more herbage, which is the result of applying our full knowledge of rain and shine, genetics and soil fertility. It is not hard to see that managing soil carbon is part of the control of what we have come to call greenhouse gases, and there is opportunity to lock up, that is, sequestrate, a lot in the soil, though we do need to realise it may be rather a one-off thing, perhaps several tonnes over a number of years. There is an equilibrium, a balance between what will be released from the soil and what might be added by certain management practices. The level is affected by climatic factors, for instance, long cold winters allow more to be held in some forms than do mild winters.

Image 25: Crop stubble, Mallee, Victoria, Australia.

6.12 Phosphate cycle

As we have noted earlier, section 2.10, P is a critical element in several processes in plant growth, most importantly in the energy transactions. We have also mentioned that P may be concentrated in the food chains of marine life, and through birds deposited on islands. We have had to complete this cycle by bringing this P back to old soils, in the southern hemisphere from Christmas Island in the Indian Ocean, and Ocean Island and Nauru in the Pacific Ocean. The vast deposit at el Jadid in Morocco owes its existence to collection in the Florida trough, and there may have been some intermediate concentration. The origins of the large deposits along the Gulf of Suez and in northern Australia are less understood. All have been an important part of the planning for maintaining P levels. The recapture after carrying of phosphorus into cities and its concentration in sewage effluent is a new and important challenge. Awareness of cycles, and care not to disrupt them, is an important part of being good human managers.

6.13 Altering the soil surface

Tillage, which disturbs the soil, has long been seen as necessary to prepare a seed bed for the crop planting, and at times to kill weeds. This has a big impact on life in the soil and in the elements that are held there, especially carbon. As a result, there has been a trend to reduce it, even down to the one pass for sowing the crop.

The simple digging stick of early humans was in due course to evolve into wooden implements pulled by domesticated animals, the oxen, donkeys, horses and camels, then more effective and longer-lasting iron ploughs and specialised implements for various activities. These include mould-board ploughs to bury residues and weeds, disc ploughs to chop residues, peg harrows to break down clods with minimal effort, implements that jumped stumps, and machines that cultivated and sowed seed at the same time. More and more often these were pulled along, and also their mechanisms driven, by engines using fossil fuel. All of these operations have implications for the life of the soil, for how nutrients are made available, for the maintenance of organic matter and then carbon levels. It may even influence soil temperature, which can affect nutrient uptake. While reducing tillage has benefited the life of the soil, it also has increased energy efficiency. Weed control, using a combination of agronomic practices, such as hay cutting, and small amounts of chemicals is far more energy efficient.

As stated above, with sowing into stubble there is minimal soil disturbance. This is part of a now very popular system, usually called conservation farming, which combines leaving all plant residues on the surface of the soil, and in many cases minimum tillage (min-till). This, unfortunately, can easily lead to crop yields spiralling down, as the extra carbon stimulates multiplication of organisms, and tying up of the other elements such as nitrogen.

So, this sequestration of carbon needs the support of science, precise analysis and modelling of the system. It also mandates knowledge of the carbon cycle, with using a carbon balance sheet. The farming system might have a period when soil capital is used up, carbon level falls, and soil capital building, when the organic matter is built up. The ley farming system used in southern Australia from the 1950s on had this, with several years of pasture building up organic matter and nitrogen level, then a period of cropping, in which these assets were exploited. Probably the optimum is to retain all possible organic matter, apply enough fertilisers, especially nitrogen to 'feed' the soil biota, and till as little as possible. Tightly monitored, this will give the lowest greenhouse gas emissions per tonne of grain produced and highest grain yield.

6.14 Get the best from crop plants with modern teams

Though we are often impressed when we see a large individual tree, say a big old apple tree loaded with fruit, in terms of maximising yield from resource inputs or land area we are more likely to do better with the available shine, rain, nutrients and land if we have other arrangements. Even with one specimen, to make it produce efficiently is a challenge, for instance, a fruit tree in a suburban garden.

Greatest productivity usually means grouping the same crop plants together as it is more economical to impose management practices on a larger group of plants. There are many opportunities to make the plants collectively more productive. Also, a range of different specialist scientists fit into the scheme of things, Table 3.

Table 3. Specialist groups which can help improve crop productivity.

Specialist type	Activity
Plant breeders and geneticists	Get a better combination of genes in the new plant
Soil biologists	Unravel and understand the complex relationships of the myriad of living things in the soil
Plant nutritionists	Make the level of nutrients available in the soil as ideal as possible for each stage of development of the growing plant
Foliage architects	Study the ideal plant shape and size and planting patterns
Bio-modellers	Set out models of the environment, especially for the rain and shine, and match these to a model of the plant, as it proceeds from inert seed to mature plant, and to determine the attainable yield
Marketers	Define the best way to care for and handle the fruit in moving it from the field to the point of sale
Nutritionists and chefs	Study cooking and preparation methods, and combination of foods

Overall, the effort is to maximise the return in terms of human need, especially for food, but sometimes for fibre, and increasingly, for useful compounds for medicinal and other uses. The term 'crop improvement' then broadly covers these efforts.

117

It is easy to see how humans can intervene to improve an annual crop plant, that is, with a generation each year, but it is much more difficult and costly to improve perennial crop plants, with a life of many years, say 20 years for a peach tree and 100 for an olive tree. Even if a better variety is developed, it is more difficult to have its use adopted. Sometimes this can be done by grafting on to the old stock, but other times it means establishing new groves or vineyards on new sites, a costly business.

We have referred earlier to the question of the advantages and disadvantages of using perennial plants for major crops, section 4.2. These include wheat, and the main attraction being not having to sow the crop each year. Annuals are geared to move as much as possible of their substance into their seed heads, to guarantee continuation of the species. In contrast, perennials put more into preserving the old plant. Then new varieties are constantly being developed for a host of reasons: disease resistance, growth efficiency, and nutritional value, that includes medicinal applications. Also, the annual changeover offers opportunities to be very adaptable. Sowing each year but with absolutely minimising effort, for example, through minimum tillage, helps give the annual very real advantages over the longer-term perennial.

When considering commercial aggregation of different forms of growth into production systems, we have to consider the reliability of rain and shine. Taking averages over periods of time, shine is fairly predictable, though the accompanying temperature of the air is less so, and the amount of rain is the least predictable. Further, in any latitude with a given climate type there can be some interesting differences; and then, for that location, substantial variations from season to season. These can have a profound influence on the hospitability of patterns to plant growth, and therefore potential for agricultural production. For example, above we compared the Californian Mediterranean zone with southern Australia, section 3.2, which have very different sites for plant growth, the former having heavy storms and a lot of run-off, the latter low rainfall per wet day, and little run-off. Though in southern Australia there are occasional heavy storms that can cause erosion of bare soil.

Soil capacity for moisture storage matters, too, and modern farmers both encourage this with management practices, and use crops which have root systems capable of using the full depth of storage. In Mediterranean climate areas humans use crops which fit an ideal rain pattern of an autumn moist enough for seedling establishment while there is plenty of shine for growth, a fairly dry winter when plants are not so actively growing, and plenty of rain in spring as shine increases so that plants grow actively, followed by a dry summer for harvest.

While the dramatic multiplication of seed grains is impressive, there has been a general increase in the ability to produce biomass, that is, plant mass, in a given combination of rain and shine. For instance, in southern Australia, many soil types were low in P, N and elements like copper and zinc, and carried sparse growth. Adding P and copper and zinc, and sowing legumes, which can capture N from the atmosphere, has given huge increases in interception of solar energy and in herbage production and livestock carrying capacity, up to 10-fold. This also given large gains in the amount of carbon in the soil. Growing so much herbage means a large intake of carbon dioxide from the biosphere, and, in the case of forage production, is some counter to the methane which is produced by grazing animals.

Managing interactions with other organisms is a challenge in all systems. Making sure the chosen plant did not suffer competition from other plants was for long the task of people, often women, with a hand hoe. As mechanisation of farm tasks proceeded, some quite clever inter-row tillers were invented, but they were never perfect. Earlier we discussed the organic chemicals, section 1.3, a vast array, with many occurring naturally in plants, or closely related to some that do. With the understanding of these came some clever use of organic chemicals that would kill one species and not affect another. These were called selective herbicides, which have now been refined and widely used. One example is glyphosate, commonly bought as Roundup, which intervenes in the plant processes, 'starving' the plant, rather than being toxic, and because of its composition would appear to have a very low probability of residual effects. So, evolving farming systems, in their quest to optimise response to rain and shine, have had various combinations of plant selection, tillage, fertilising and control of weeds.

Take away messages

1. Agriculture, which includes crops and grazing animals and fruit and vegetables, can be seen as an ecosystem with a maze of interactions, which are heavily dependent on the solar powered green plants.

2. A key challenge is to have plants that 'make the best' of shine and rain.

Shine

3. We maximise shine interception by orientation of plants, for example, in running rows north-south, and pruning.

Rain

4. To water plants in dryer areas irrigation systems combined with water storage may be used.

5. Modern farmers manage the capture of rain and optimising soil capacity for moisture storage, based on science, and use a computer model to incorporate this into management practices.

6. In dryer areas it can be important to manage the movement of salt in the landscape, to minimise reduced crop growth due to high salt levels.

What's needed in soil

7. There must be an adequate and available supply of all of the essential chemical nutrients for plant growth from the time of germination to the ripening of the seed.

8. It is optimum to retain all possible organic matter, apply enough fertilisers, especially nitrogen to 'feed' the soil biota, and till as little as possible. Tightly monitored, this will give the lowest greenhouse gas emissions per tonne of grain produced and highest grain yield.

9. It is important to understand nutrient cycles, in particular for C, N and P.

10. Plant nutrients, in particular N and P, should be conserved for both sustainability and minimise nutrient runoff.

11. Organic matter applied to soil must have the right ratios of critical elements such as carbon, nitrogen, phosphorus and sulphur.

Modern farm practice

12. Gain in yield in any district, that is assessed to fit the trend line to long-term yield data, is a part of finding the right cultivar and using the right farming practices that can cushion the risks.

13. Modern practice to seed fields uses narrow seeder points, often with the machine guided by a GPS, to sow the next crop in between the rows of the last one, with minimal soil disturbance.

14. A science-based approach to optimise crop growth involved collaboration between a diverse range of experts.

7

Producing crops under rain and shine

7.1 The challenge in producing crops

It is useful to see modern agricultural production, which is very much high tech, against our understanding of rain and shine. The aim is to fit a number of a crop plants grown together as neatly as possible into the rain and shine environment, with other plants attendant, and insects and weeds managed effectively to allow maximum saleable products of high value from the crop plants. As there are more and more people on the planet, greater efficiency, usually expressed as yield per unit area, or per unit of energy, has been sought, with never ending evolution of farming systems as new technologies are developed.

7.2 The human drive for improvements

Even in a general sense it is difficult to believe hunter-gatherers were not intelligent enough to be a bit manipulative, perceiving advantages, encouraging more useful plants in more convenient locations, for themselves or the animals they hunted, to have some system.

From that situation crop production has developed through many stages, responding to, and aided by knowing more science and having not only better digging sticks, but a wide range of other technologies. The better 'sticks' were soon drawn through the soil by draft animals, then by engines using fossil fuels. Specialised 'diggers' like the stump-jump plough, were developed, especially to deal with the newly cleared land of southern Australia. Collecting the fruit, especially if a grain, was in due course by stripping the heads of wheat and other crops, and cleaning them separately. Again, Australia led in this with the Ridley wheat stripper. Then came the combine harvester, with a clean, finished product and today, even a map of the yield of the field.

Weeds were cut back or pulled out by hand, then hoed, by a specialised digging stick, and disturbed in the soil by a cultivator of many tines. Finally, to avoid the high energy cost and physical damage to the soil, by passing over the land with a light implement delivering a chemical spray. This enabled past crop residues to remain on the soil, and the living soil to remain undisturbed. The digging stick is not used much in truly modern agriculture!

For annual crops there was always the possibility of changing crops to ease the effect on the land, or to avoid disease, called crop rotation, or even using different land every few years, as in the slash and burn. Fertilising was tied to life and life cycles. Some communities continued return of human wastes, others were more fearful of disease. Wonderful examples are given in books about Chinese farming, for example, 'Farming for 40 Centuries'. In cold climates animals were protected in winter, literally sharing the building with the humans, and their waste and bedding was returned to the fields. Where there was no winter shedding there was little or no return, and unless the soil was naturally very fertile, yields declined. Resting and giving some time for natural breakdown of parent rock material was the basis of the slash and burn.

Selection of plants was practised. Early farmers wisely collected seed from their most successful plants, implicitly selecting for some disease and pest resistance.

7.3 The on-going search for improvement

Production systems have always combined the above components with varying emphases, reflecting available technologies, available resources, especially labour, and contemporary social values and prejudices. It is important to realise, however, that the production systems are much more similar to each other than they are different, each in its own way dealing with the interactions of rain and shine and the resultant issues raised above. It is only realistic to admit that they have had down-sides as well as upsides. What is important is the early recognition and acknowledgement of this and the on-going search for improvement, so not settling into a 'traditional' pattern. Also, so-called traditional farmers in developing countries have down-side effects, just like modern farmers in Europe and Australia.

Some consider farming, the very deliberate cultivation of plants and the domestication of animals, as the beginning of the end for planet earth, as the onset of severe, unsustainable impact of human production processes, and, with its apparent success, the development of specialist activities and hugely destructive cities. They consider the hunting and gathering stage was sustainable long term, but modern farming unsustainable and destined to collapse.

Perhaps we should look back and ask the question "Was there a definite point when hunter-gathering gave way to farming, as a point of no return for some people, or was the former simply part of this evolution in management?" This is important in answering those critics of farming who see it as our nemesis. Did hunter-gatherers become the first farmers? Surely they would have long been seeking better tasting, and possibly more fruitful plants, even plants with insect and disease resistance. One cannot believe intelligent beings would have not done so. Once discovered, they would have gone back to these each season, even encouraged growth by simple water diversion. And did they learn to prune out the competing plants from their elders? If they weren't farming by gathering seeds and relocating a few seeds, at what point would they have been doing so, 10 seeds or 100 seeds? Was it when they pulled out the competing plant instead of just pruning it? And would a person who learned game animal behaviour, when they came to drink, which feed they liked at dawn, and avoided killing pregnant females, be simply a hunter, or be an animal farmer, the forerunner of our modern range-land grazier? Where does opportunistic irrigation from floods, or simply terracing a little around existing plants, fit in?

It is easy, looking back over many millennia, to fall into the trap of considering a change that crept in over centuries as sudden, as in cropping spreading from northern Greece into the rest of Europe, or from along the mighty Euphrates along the Grand Silk Road to China. Again, we need to be reminded of the size of the human population. Early, numbers grew slowly because of periodic disease, especially of the young. We, today, live in a world where hygiene and medicine, stimulated by the same sciences as agriculture, play a part in increasing pressure on resources.

Though the total effect is one of slow evolution of systems, inventions have sometimes enabled leaps, just as in our time we have seen major technologies applied, like the computer.

Take away messages

1. In modern agricultural production the aim is to fit a number of a crop plants grown together as neatly as possible into the rain and shine environment, with other plants attendant, and insects and weeds managed effectively. This is to allow maximum saleable products of high value from the crop plants.

2. There has been searches for improvements in mainstream agricultural systems, yet there have been down sides as well as up sides.

3. What is important is the early recognition and acknowledgement of issues and the on-going search for improvement, so not settling into a 'traditional' pattern.

8

From gardening to farming

8.1 Moving to more complex systems

Against a background of a less than rosy life as a hunter-gatherer, it is not too difficult to imagine people gradually doing things that made the system more reliable, with a more certain and greater return from their personal exertion. For instance, to group plants for easier management, separate trees from vegetables, and grazing animals separate from grain crops. At many different places, rather than one group specifically 'inventing' it, we can imagine many different people recognising in a general way what we can now state explicitly. That is, some of the advantage may lie in more effectively capturing shine, some may be with spreading the water of the soil, some to do with fertilising at the right time, some to do with easier harvesting, and in due course using tools, even simple machinery. It is suggested that the famous Ridley wheat grain stripper invented in South Australia in 1844 was based on a machine mounted on a chariot chassis that the Romans used in Gaul in 100 BC!

It is generally accepted that early locations for farming included the fertile crescent in Eurasia, along the Euphrates and Orentes rivers and into Egypt, and the northern plains of Greece. It is no accident that these were areas with an expanse of fertile soils. Though, many argue that similar activities were carried out in east Asia and the Americas. We have already noted that riverine plains and deltas had their fertility 'recharged' by floods, section 2.7, and of course the mixed deposits of small particles of rocks would break down and release plant nutrients reasonably quickly. What would have been the total impact of these first farmers on the environment? Is it better to 'farm' a small area for a large amount of food, or more lightly impact on a much larger area, as the gatherers did? To some extent the answer must relate to population. With more people to be fed, there would be advantages in concentrating the growing, and getting a higher return for all inputs, including human effort. The population was apparently growing, slowly, but nothing as rapidly like it has in the last century.

Return of wastes would have partly countered the removal of nutrients, and with modest yields at favoured sites where fertility was added from parent material breakdown or floods, there may have been long term maintenance of the soil nutrient levels, that is, it may have been a sustainable system. As the society became more complex, with consumption at a distance from growing the crop, removal and relocation of nutrients has become an increasing problem, peaking in the large sewerage systems of present-day metropolitan areas. For animal grazing, camping grounds and feedlot systems cause considerable relocation and concentration.

Small areas of production we tend to call a garden. Can we model large field growing areas on it? Characteristically, gardens have a range of crops, as are polycultures, and attain high yields per unit area, but have a high input of human effort per unit of production. Organic matter is treasured and conserved, normally added to by collection of material from outside, such as animal manure from animal camps or pens. Nowadays there are so many sale boards for cow poo or horse manure on roadsides! A garden is generally heavily reliant on inputs from other areas or sources.

Gardens also have such a mixture that the question of co-cropping is relevant. However, this is a convenience in a small area, and with larger areas equal numbers of two crop plants will give a greater yield in two separate plots or fields than if mixed, even in inter-row patterns. In many areas mixed farms evolved: animals for milk and skins or wool, poultry, pigs to consume food wastes, and even rabbits. Animal wastes went back to crop land. Some people long for these to return as the way to feed the world, but any critical analysis shows efficiency of energy use and other resources is lower than in specialised production.

8.2 Attitudes to different farming systems

The term conventional agriculture is frequently used interchangeably with traditional agriculture. Though, obviously this is unfortunate, as it implies that mainstream agriculture is not innovative. which we have shown is incorrect. Yet, these terms are often used by advocates of specialised, philosophy-based systems, for alternative systems such as organic farming, or biodynamic farming, or permaculture, to differentiate their system and activities from the widely used ones, and sometimes to perceive their product as an advantage. These other systems usually have some special objectives and may set those over and above the maximising use of shine, efficiency of water and energy use and maintenance nutrients level in the soil. Also, the word chemical is used more as an epithet, but all life needs chemicals.

In Western countries there are also increasing numbers of what are best called life-style farmers, who have a different attitude to certain aspects of mainstream commercial farming. Some are poor, but many are moderately wealthy, having non-agricultural sources of income, owning land and having some farming activity as an interest, but for whom maximising use of shine, rain and soil nutrients, and so making a profit, is not a prime objective. Though they produce saleable products, the task of feeding the world cannot rest on their shoulders, any more than it can on the peasant farmers. With improvement in transport and communications, life-style farmers may live quite a distance from major cities, may be a professional person working largely on the internet, and/or flying to the city for a couple of days a week. So, their presence affects many rural communities, positively in their economic and intellectual contribution, and negatively in the dilution of the so-called traditional values of rural society. Some popular authors allege that the change of the community from simple agrarian life signals ecological collapse, but such arguments do not stand scrutiny.

Conventional is also a descriptor used for the agriculture of yesterday, of the last decade or so, for the land managers who have not yet adopted the leading-edge practices. As is the nature of take-up of new things, be it the motor car or e-mail, there are always some people using a horse and buggy or snail mail, and those equivalent in agriculture are described as conventional.

Yet, mainstream farming systems are continually evolving, to adopt new practices being generated from scientific research and inventions. We now look at the different types of agricultural systems, how well these fit into this innovative approach, and the sustainability of these.

8.3 Traditional agriculture

This term is often used but what does it mean in terms of the basics of making plants grow? It has been defined as farming based on the long-used practices, things that have been part of farming for centuries. That is, deep human involvement in processes, so less mechanisation, minimal use of chemical solutions to control pests, so, often lower yields per unit area, less use of so-called artificial fertilisers, so, at best steady yields, but often declining yields. There is even a negative attitude to science and scientific knowledge. Many people also associate it with lifestyle, with simple folk sitting contentedly at their hearth, eating what they grow, being part of stable families living a wholesome, healthy life.

Central Africa provides some answers to the question as to whether there has been a set pattern and has remained unchanged for hundreds, perhaps thousands, of years. Typically, at one time there would only have been a few people with a small range of crop plants. Yet, today we can make a long list of plants being grown, most of them originating elsewhere and now grown throughout the tropics. These include; cassava, haricot beans, soya beans, groundnuts (peanuts), taro, cabbages, bananas, avocados, mangos, pineapples, sweet potato, tomatoes, sugar cane, sorghum, coffee, lemons, and quinine. Tools have changed too. Once there was no iron for hoes, and now we can only wonder when the now widely used heavy iron hoe was first used.

These changes suggest that there is no question of clinging to any fixed system. The challenge is rather to continue the tradition of incorporating new things, sensibly using our modern-day advantage over our ancestors in that our communications and network can bring us new things, and against this we must test and select. Examples of feasible improvements for Central Africa are new varieties of present crops with disease resistance, new lightweight steel hoes, concentrated fertilisers, for instance, to remedy deficiencies in P, but not necessarily mechanisation and a great range of agricultural chemicals.

8.4 Peasant farming

This descriptor applies to quite a variety of systems, partly because different crops are used. In Asian countries it has been mostly rice based, and irrigated for the most part. In the Americas it has been corn, as called maize, based, normally on rain-fed, in northern Africa, for wheat. The common elements have been low level of capital investment, little mechanisation, high proportion of human physical effort, low inputs of chemicals, poor yields, and the absence of scientific support.

In peasant farming unit areas of cultivation are small, increasingly so if land inheritance is not amended by land reform, and then fragmented. There is little crop protection, yields vary and because the farming is subsistence for that year, with little storage, so communities are easily threatened by famine.

The slash-and-burn system is a variant of peasant farming, usually used for corn production, see Image 26. The burn releases some nutrients, and the spell of a few years, even up to 30 years, from farming, allows some accumulation of nutrients from parent material and organic matter break down. Speeding up the cycle as food demands increase is a very temporary solution, often leading to soil degradation.

Image 26: Slash-and-burn system, with the ever search for fresh land, location south of Rio de Janeiro, Brazil.

131

Variants of these systems, even today, are common in so-called developing countries, frequently causing the people to live at starvation levels. Yet some people still glorify them as the noble efforts of indigenous people, longstanding, sustainable, using traditional varieties and to laud the system for not using 'chemicals'. Yet, in winter the sparse crop captures a low percentage of the shine, and the crop has a poor root system to access mineral nutrients and deeper soil moisture in spring. Diseases and insects attack the crop. All too often the genotypes used do not mature before the rains cease, giving a poor and disheartening return for effort.

Critically, yield per unit area from such farming has not increased over the last century. In past centuries when more food was needed early farmers used natural regeneration of forest to renew fertility, or collected and applied organic remains, that is, crop residues and animal manures. Alternatively, new land was brought into production, either locally, by ploughing up grassland or clearing forest, or speeding up a slash and burn rotation, or by invading new areas, even new continents previously lightly occupied by hunter-gatherers. It is only in recent decades that society has resisted this raw expansionist approach, as we recognise the value of forests such as the Amazon for sequestration of carbon dioxide, and demand the setting aside of land for wilderness areas, National Parks and other recreational purposes.

Then, the only answer to greater demand for food is to gain greater productivity from the existing farming areas. The move from draught animals to tractors deferred this conflict. In the USA the tractor freed up 40 m ha of land, nearly a third of the total arable land. Organic production works best if residues are collected widely and concentrated on small areas. This scheme was used successfully for many centuries in medieval Europe, where dung was an important enough commodity to be recorded in Parish registers. Early farmers collected seed from the most successful plants to gradually improve their harvests, and much more recently, cultivars with improved yield, quality, and pest and disease resistance, are being bred with increased precision as the science of genetics has further developed. Also crops can be effectively grown on steep lands, Image 6, as well as normal horizontal lands.

8.5 Alternative systems

Though the word alternative is used, in fact these have more similarities than differences when one considers scale and type of product. Small areas of horticultural production are frequently established on deep, friable soils with a large reserve of nutrients, and in the short term need little fertiliser. When they do, organic residues, rather than manufactured fertilisers, are often used, and it is practical to do so for modest areas, as the quantity of materials is limited. One example farm is sited on deep (20 m) fertile red basaltic soils and used Dynamic Lifter, composted chicken manure, and Guano Gold, a bird refuge residue. Weeds were controlled by using hot water.

Those attempting organic operations over large area cereal production have a battle maintaining phosphorus levels. Rock phosphate, that is, mined and not processed, is moderately successful on very acid soils in rainfall above 700 mm, so with soil moist for much of the year. Yet, it will take time to be effective, often several years. The option of adding more sulphur, to make the soil more acid is not attractive. Instead, lime should be added to make the soil less acid. For cropping areas rainfall is usually less than 600 mm and rock phosphate in ineffective. Again, alternatives, such as crushed coral can be used for limited areas, but the product is too limited in quantity to be adopted in broad-scale farming.

Yet, some fascinating local area farming systems occur. One is the Dehesa of western Spain, where pigs and sheep are run on grassland under an oak woodland. As the pigs eat acorns, the meat is oak-flavoured and so sought after as a delicacy.

8.6 Organic production

In organic production the emphasis is on a system of inputs that, it is believed, will result in produce with comparatively better food quality, with the absence of significant chemical residues, and, it is boasted, better taste. Typically, it is labelled 'safer, more nutritious and better for you'. Few would quarrel with these objectives, to provide tasty food free of nasties that can reduce health. The very real benefit of organic concepts to society over recent decades has been to provide a stimulus for more attention to be given to nutritional value and freedom from residues. These objectives, then, have been incorporated in mainstream agriculture.

Currently there are alternative guarantees, such as Good Agricultural Practice in Europe, and stringent analysis to ensure low levels of nasty chemicals. Even the tendency to monopolies in supermarkets has helped, as these have high standards built into their supply contracts. A number of recent comprehensive surveys have reported no detectable differences between the standard of foods from organic farms and those from mainstream production.

To a fair extent the organic producers can only continue to thrive by being able to contrast their produce with those not so labelled, setting up an 'us' who are righteous, and 'them' who are baddies. Most often the contrast is subjective, not based on analysis of either group of products, but on management systems and unprovable assertions such as: 'It tastes better' or 'It looks better'.

However, in considering the rain, shine and soil nutrients criteria, the first important thing to recognise is that yields are often lower than for the mainstream food sector. From the producers' standpoint, higher prices make up for this lower efficiency.

Currently organic agriculture occupies about 0.35% of agricultural land, mostly in developed countries, and has disproportionate visibility in markets. Yet, if the systems were adopted throughout, more land would have to be brought into production, and more energy used. At 80% of mainstream yield, 25% more land would be needed, at 66%, 50% more land. Also, where land is limited this could cause conflicts over forests and public land. The same applies to energy use, even including allowance for the manufacture of the chemicals used by the mainstream. This is because there are more emissions of greenhouse gases from organic production. Importantly, it is only while most producers use the rain, shine and soil nutrients to maximise production that a minority of the populace can indulge in the preference for organic farming. A recent review in the journal Field Crops Research concluded, that using the organic system on the existing land area, enough food would be produced for just 3-4 billion people, well below the current population of 7.5 billion and far below the estimated 9.8 billion by 2050.

Secondly, much organic production is not sustainable in the very long run. This production generally occurs in small areas, and much of the production is of specialised crops and fruit and vegetables. In these cases extra organic materials, such as animal wastes, are frequently brought in from outside. So, the areas where the animals graze may suffer nutrient depletion, or if it is from intensive units, back up the chain where the fodder is produced. Many intensive units are only themselves sustainable if some other activity guarantees to use their wastes.

Many producers are located on deep, fertile soils with either some continuing release of nutrients from parent rock, or such a high bank of nutrients that a slight decline may be imperceptible in the medium term. So, our judgment must depend on whether we believe sustainability is a system for a thousand years, as if we intended to farm forever.

In contrast, on poorer soils, with little natural fertility, nutrients must be added to replace those removed in the crop. Supplying these in bulky organic materials involves high freight inputs. That is why broad area farmers so often opt for concentrated fertilisers. Overall, for those caring about sustainability it may be better to insist on a statement of what nutrients have been added to the soil, as well as what has not. Hydroponics, for which production occurs on almost pure sand, is the most likely agriculture to be completely sustainable, as the nutrients must be added before a crop is grown.

Organic agriculture fits quite well into the total system of production in developed countries, especially those with a cold winter. Some organic residues are thus available, and must be disposed of, from the housing of animals in winter. Then, huge amounts are generated by intensive animal industries and feedlots. In contrast, in developing countries there are no such sources. Material available is often bulky, difficult to transport, slow to break down and very low in essential nutrients.

In summary, the organic movement has been successful in influencing all food production to manage herbicide and pesticide chemicals carefully. Yet, there are very real problems in sustainable production as expanded organic farming need to procure adequate nutrients. As a result, many organic operations have widened their acceptance of mineral fertilisers, as replacement materials.

8.7 Biodynamic farming

Closely related to organic farming we have Biological farming, which in Australia is supported by the association called Biological Farmers of Australia. This is a bit cheeky as all farmers are by definition involved in biology! In the same sense all growth is biodynamic, but a particular group use the name for their system of growing plants, including agriculture and horticulture, which developed from the teachings of Rudolph Steiner through his lectures in 1924. He argued that the health of soil, plants and animals depended on (re) connecting nature with the creative forces of the cosmos. He described ways of doing this, including homeopathic preparations. It is argued that instead of the 'chemical' approach, where elements like N and P and K are added as chemicals, these can be made available through living processes, such as liquid brews and compost heaps. In practice this has much in common with organic farming, with the addition of some specialised inputs and timing of operations to fit lunar cycles. Certification of farming operations as Biodynamic is done through the Australian Certified Organic Organisation. Also, like organic farming, it sets itself as a contrast to 'conventional' production. Again, such a system is unlikely to have wide application in feeding the world's billions of people. Though the prominence given to this movement assists in support for those commercial producers who have a deep concern about chemical residues.

8.8 Permaculture

Permaculture, too, has much in common with organic production, and many laudable emphases. One is the use of rain and shine falling on a particular piece of land at a particular time being the limits to those inputs, rather than using resources from elsewhere, at times even impoverishing other land. However, in its purest form it is only practicable on good soils in high rainfall areas using a good deal of capital. For example, in setting up alternative energy systems and storing a good deal of water.

In practice one usually finds that the boundaries are breached, as there will be mains power, connection to an external source of water, even just a creek running through, and procurement of additional nutrients, by bringing in animal manure and other organic residues. It is more applicable to intensive production such as fruit and vegetables than broad-scale farming.

8.9 Science-based agriculture

In contrast to the alternative systems, it can be confidently asserted that on the mainstream agricultural land areas, production based on our awareness of science for rain and shine and genetics and nutrition could support the global population in the long term. Bringing these factors together in a unified knowledge base is the essence of scientific farming. Though it has 'flowered' in the last few decades, it is a creature of the last two or three centuries. It is knowledge based: Greek scienta means knowledge. It also includes testing hypotheses with experiments, either on the farm or at a nearby research station, and evidence is collected and analysed. Its basic concerns have been maximising the use of rain, shine and nutrients, with recent use of modelling to integrate and assist in decision making. It has recognised, and been an important part of resource management and its emphasis on sustainability.

Its origins are in a broad sense traceable to Europe, especially Britain, explaining the gains that had been made from improved rotations. European farmers had relied on animal manure to maintain the level of nutrients in the soil, from animals that were grazed on other lands, often commons and sheep walks on Manors and Castle estates, and often with some legumes, which helped capture nitrogen. The Romans actually promoted legumes as making the soil more fertile, but that had been largely forgotten.

Most of Europe, and especially Britain, had very favourable rain, enough each year, and well spread through the year. While shine was not often limiting for growth, temperatures were too low for growth, and animals had to be kept inside out of the weather in the winter. Root crops such as turnips and beet, with a rosette foliage, were very good at intercepting weak shine and building up a lot of sugars and starches. These were incorporated into rotations, harvested and fed to the animals in winter, and the 'muck' put back on the land, so returning a substantial proportion of the nutrients taken in the harvest, though not all. The rotations also included legumes, assisting the maintenance of nitrogen levels.

This organic farming system more or less maintained yields of grain crops grown for human food over many decades. However, as the population grew, slowly, as being sometimes depressed by the plagues and other pestilences, famine was not uncommon in Europe, even into the early 1800s. There was some fallow land, that is, land that was out of crop sequence, resting, and used for animal grazing. At this time nutrients were being processed by soil micro-organisms, and made more available for plants. The term bare fallow, that is, land kept nearly free of any growth, is more recent.

At the same time there was a spread of cereal farming to the New World, that is, the Americas, and, a little later, Australia. There, old fertility, that had accumulated over centuries, was at first able to give satisfactory yields, though yields soon declined on some soils. Bare fallow was used to arrest this, but the effect was only temporary, and much land became exhausted and eroded.

In the 1850s in Britain experiments were conducted in adding grounded bones, to provide phosphorus and calcium as well as a liming effect, and bird guano that mainly provided phosphorus, which together gave large yield responses. With hindsight, we realise that however thorough the organic farming system was in applying muck and leaves, quite a lot of P was being taken away in the grain sold to town people, and not returned. Not enough was being released by breakdown of parent material to make up the deficiency. So, this most vital of plant nutrients, the key to capturing shine, was running down, and there was a downward trend in yields on many soils.

The site for the long-term experiments at Rothamsted in the UK, concerning crops and pastures, now carried on through 150 years, is one of the great 'shrines' for agricultural scientists. This was a defining time in humans' history of making plants grow, marking the beginning of understanding of plant growth efficiency. The tests both confirmed the need to remedy the low levels and studied how to then maintain them. Soon other nutrient needs were defined, and, as an outcome, today's precision farming is strongly knowledge based. Precisely managed crops are shown in Images 3, 4, 21 and 24.

For extensive crop areas in southern Australia, over the last 50 years, the generally adopted method has been to use processed guano, called superphosphate, and natural organic restitution of the soil, using legumes, in symbiosis with bacteria, to build up soil nitrogen and organic matter. The key to success has been the use of phosphate. Increasingly, people put a lot of effort into the land and invested capital in buildings and fences, but also, not widely recognised, poured very real capital into the soil. This included fertilisers, in putting P into the soil bank, land levelling, and clearing stumps and stones. Private ownership of land, or at least long lease, became part of life and an incentive to improve land.

The importance of using legumes to access 'natural' nitrogen from the atmosphere cannot be overstressed, even if there is some topping up at critical times with manufactured fertiliser. Development of agriculture in southern Australia has been a great success story for use of legumes, to improve pastures, sometimes later in rotation with crops and legume crops, Image 3. This biotechnology has been successfully exported to other countries with a similar climate, for instance, Israel.

The modern farmer, seeking to maximise the value, to reach what is described as the optimal attainable yield, is likely to have a system that includes a number of key components, Table 4.

Table 4. Key items in a sustainable system to optimise yield.

Item	Properties
Choice of seeds of varieties with all possible available genes	For seedling vigour, including resistance to soil borne diseases; for resistance to leaf and stem rusts and the like, to minimise use of any chemicals for disease and pest control; for resistance to frost damage at flowering time; for the best possible protein and other nutritional attributes
Understanding of foliage architecture	To capture maximum shine: shape of shrubs and trees, direction of rows, grazing to prevent pastures becoming too leafy, by the concept of Leaf Area Index.
Minimal soil disturbance and physical damage	Minimum or even zero tillage, so maximising retention of carbon, minimum expenditure of energy and minimal soil disturbance, enhancing soil physical attributes and minimal effect on other living things in the soil
Alternatives to the use of tillage to control weeds	Use of small applications of 'smart' chemicals, alternative crops to break cycles
Regular measurement of soil features and addition of fertilisers	To both maximise yield and fully replace the nutrients taken out by the crop; adding some nutrients in pre-crop activity (such as N from pastures or legume crops) some at seeding time, and some later in growth as foliar applications sprayed on at the perfect time of need, such as grain formation
Timing of seeding and innate growth phases	To mesh with the expected rain and shine effects (as predicted by sophisticated climatic models)
Equipment to sow the maximum area	To what is deemed the ideal time and conditions

The overall aim is to have a crop that fully utilises shine, light energy, falling on it, especially in winter and also rain, especially in spring and early summer. Crop modellers will have determined the attainable yield, based on the shine at that place and with adequate soil moisture. It is a fundamental that nutrient deficiencies do not limit the yield, but that there is no expenditure on additional fertiliser if moisture is limiting. So, fertilizer is often applied progressively through crop growth, to match uptake requirements, as determined by shine and rain.

This attainable yield should be achieved in hydroponic systems, where plants are grown in nutrient solution that can be continuously monitored, but is more difficult in broad acre farming where moisture may become limiting and/or cost and access are added difficulties. A common solution is to add fertilizer in two or three doses and rely on the nutrient-holding capacity of the organic and mineral phases of the soil to provide storage and release. Soil characteristics are, therefore, important in dictating not only what fertilizers are required but also how they are applied. Sandy soils have little nutrient holding capacity and applied nutrients are easily lost.

There are debates about the form the fertilisers should be in. Some advocate only organic forms, and in the garden, at small scale, this is practicable. However, over large areas, where freight and spreading would take a lot of energy, use of concentrated fertilisers is usually optimally efficient. In any case, we must remember that only inorganic chemical compounds go into solution and are taken up by plants, so organic fertilisers must first be broken down into an inorganic chemical form.

Integrating such knowledge of shine, moisture and nutrients is now international. One company, which has greenhouses to produce tomatoes in the Netherlands, USA and Saudi Arabia has identified Gyra, in northern New South Wales, Australia, as an ideal site for a 5 ha size green house, with possibly four more later. At an altitude of 1300 m Gyra has plenty of water, has minimal cloud and intense winter shine, and clean air, as well away from cities or industry. There it is also cheaper to heat the glass houses in winter than to cool them in summer. Elsewhere, quality of fruit is often poor because of overheating in summer.

8.10 Toward sustainable agriculture

Clearly a scientific, evidence-based, approach is optimal to support sustainability of agriculture, to feed a growing global population. Yet, it is only realistic to admit that most human activities, including most aspects of farming, have had down-sides as well as up-sides. Yes, all farming systems we have described above, even so-called traditional farms in developing countries, have down-side effects, just like modern farms in Europe and Australia.

What is important is the acknowledgement of this, as well as the on-going search for improvement. That is, not settling into a 'traditional' or 'conventional' pattern. In this context, a reasonable definition of sustainable agriculture is agriculture to which society has an ongoing commitment to provide enough resources to monitor the system, research its issues, and ensure fixing these.

Take away messages

1. Understanding the sustainability of mainstream agricultural production is affected by contrary attitudes of advocates for alternative smaller scale systems.

2. Other agricultural systems include; traditional, peasant, organic, biodynamic and permaculture.

3. The organic movement has been successful in influencing all food production to manage herbicide and pesticide chemicals carefully.

4. However, for organic production, yields are often lower than for the mainstream food sector. From the producers' standpoint, higher prices make up for this lower efficiency.

5. In contrast to the alternative systems, it can be confidently asserted that on the mainstream agricultural land areas, production based on our awareness of science for rain and shine and genetics and nutrition could support the global population in the long term.

6. All agricultural systems have shown down-sides as well as up-sides.

7. The on-going search for improvement includes recognising any down-sides.

8. In this context, a reasonable definition of sustainable agriculture is agriculture

to which society has an ongoing commitment to provide enough resources to monitor the system, research its issues, and ensure fixing these.

Acknowledgements

Much of the material in this book has been derived from the Dr David Smith's teaching materials over many years, in turn gleaned from many different authors. We are especially grateful for the diagrams, accumulated from various authors, in this or some similar form, as powerful aids to teaching 'How Plants Grow'.

Weather data for Australian sites, Nagambie (Victoria), Perth (Western Australia), and Roseworthy (South Australia) were obtained from the Bureau of Meteorology, Australian Government. Data for Valencia, Spain, was from the Agencia Estatal de Meteorología, Ministry of Environment; for Los Angeles and San Francisco, USA, from the National Oceanic and Atmospheric Administration, US Department of Commerce; for Porto, Portugal, from the Instituto Português do Mar e da Atmosfera; and for Cape Town, South Africa, from the World Meteorological Organization, United Nations Economic and Social Council.

Author biographies

The late Dr David Smith AM had a long and interesting association with plants, from boyhood on a farm and in the bush, through teaching at all levels, researching many aspects, then, as Director General of Agriculture managing the resources of a large department dedicated to helping farmers grow plants better.

Dr Duncan Rouch is an independent scientist and planner, and has over 25 years experience as a professional research scientist. His interests include addressing environmental and agricultural issues. As well has having worked at universities in both the UK and Australia he also has experience in government and industry. He has over 70 research publications, in scientific journals and books.

Related book by the authors

Duncan Rouch and David Smith, with Andrew Ball, have also authored the related book, 'Saving Planet Earth. Why Agriculture and Industry Must be Part of the Solution', ISBN 9781925501698, also available from Connor Court Publishing.

www.ingramcontent.com/pod-product-compliance
Lightning Source LLC
Chambersburg PA
CBHW080557220326

41599CB00032B/6518